Oussama Jarrousse

Modified Mass-Spring System
for Physically Based Deformation Modeling

Vol. 14
Karlsruhe Transactions on Biomedical Engineering

Editor:
Karlsruhe Institute of Technology
Institute of Biomedical Engineering

Eine Übersicht über alle bisher in dieser Schriftenreihe erschienenen Bände
finden Sie am Ende des Buchs.

Modified Mass-Spring System for Physically Based Deformation Modeling

by
Oussama Jarrousse

Dissertation, Karlsruher Institut für Technologie
Fakultät für Elektrotechnik und Informationstechnik, 2011

Impressum

Karlsruher Institut für Technologie (KIT)
KIT Scientific Publishing
Straße am Forum 2
D-76131 Karlsruhe
www.ksp.kit.edu

KIT – Universität des Landes Baden-Württemberg und nationales
Forschungszentrum in der Helmholtz-Gemeinschaft

KIT Scientific Publishing 2012
Print on Demand

ISSN: 1864-5933
ISBN: 978-3-86644-742-4

Modified Mass-Spring System
for Physically Based Deformation Modeling

Zur Erlangung des akademischen Grades eines

DOKTOR-INGENIEURS

an der Fakultät für

Elektrotechnik und Informationstechnik

des Karlsruher Instituts für Technologie (KIT)

genehmigte

DISSERTATION

von

Dipl.-Ing. Oussama Jarrousse

geb. in Damaskus, Syrien

Tag der mündlichen Prüfung: 19. Mai 2011
Referent: Prof. Dr. rer. nat. Olaf Dössel
Korreferent: Prof. Dr. Sasha Panfilov

Acknowledgements

Firstly, I would like to express my deepest gratitude to Prof. Dössel, for giving me the opportunity to make part of his team, for his tolerant supervision and for his persistent encouragement.

Secondly, I want to thank Prof. Dr. Johannes Bienlein, to whom I will always be grateful, for putting his confidence in me back in Damascus 2005 and for his direct and active involvement in giving me the chance of accessing the world of scientific research, something I always longed for.

I also want to thank the Katholischer Akademischer Ausländer-Dienst (KAAD) not only for granting the scholarship, but also for all the care and support they showed and also for the various interesting and educative seminars and programs they offered. Special thanks goes to Herr Landsberg and Frau Dr. Christina Pfestroff for their outstanding dedication for the message of the KAAD despite all the difficulties.

Many thanks go to my colleges who made the office atmosphere very pleasant. Especially, Thomas with whom I shared long nights working late at the office, David, Walther, Tobias, Frank, Martin, Sebastian, Mathias, Julia and all the other I didn't mention here, who I was very lucky to have met and worked with. I also would like to thank Herr Schroll, Herr Halama and Ramona with whom I had long conversations. Many thanks to Matthias Reumann and Gunnar Seemann for their advices and for helping me adapt to the scientific work at IBT and helped me understand the IBT source code during the early phases of the PhD.

I would also like to thank all my friends who kept pushing me further, lifted me up when I was down, and supported me all during all the phases of the research. Special thanks goes to Marieanne, AlMouthanna, Ayham, Nadia, Paola, Nidal, Yara, Elias, Meike, René, Cornelia, Thomas, Sami, Noha, Nina and Oki, Kathi, Friederike and Christine. Also Amer, Laial and Lucas, who took good care of me in the hardest of times, and of course all the others. You are actually everything I am and everything I have.

Additionally, I want to dedicate Chapter 3 to Sarah McBrian, a dear friend of mine, who passed away after a brave fight with lung cancer. She will be always present in my memory.

Finally, I want to express my deepest appreciation towards my father Nidal, who insists that I should be calling him my friend, and Kawssar my beautiful loving mother, who was always at my side and in my heart. Without their influence, support and sacrifices, I wouldn't be able to understand parental love, nor be able to follow my dreams.

Preface

In this work a physically based method for modeling elastomechanics and several implementations of the method are presented. This was mainly motivated by the search for an efficient method for personalized computational modeling of the human heart mechanics.

Besides Chapter 1 which contains a general introduction, and Chapter 6 which presents an outlook, the thesis can be separated in two main parts.

The first part presents the method developed for physically based modeling of elastomechanics in details in Chapter 2. The method developed in the course of this work is based on mass-spring systems, known for their simplicity, computational time efficiency and their ability to handle large deformations easily. It improves upon ordinary mass-spring systems by bridging the gap between the constitutive laws known from the continuum mechanics and the simple implementation of mass-spring system removing the need for spring parameterization. It also adds the ability to control anisotropy, impose volume preservation and inlcudes various boundary conditions.

The second part presents several implementations of the presented mechanical modeling method. These implementations cover modeling breast mechanics in Chapter 3, ventricles deformation in Chapter 4 and using the model of elastomechanics in 3D elastic image registration in Chapter 5.

It was intended that each of the mentioned chapters starts with a motivation, followed by an introductory section, then the methods used or developed, and finally a section detailing conducted simulations and the related results before the chapter concludes with a discussion related only to the subject of the chapter.

This choice was made in favor of the traditional scheme, of "state of the art" followed by "new methods" and "results", to avoid disorienting the reader with the broad span of subjects presented in this work, and to center the focus on the methods developed for mechanical modeling. In several cases the methods described in one of these section contained both, methods implemented from the literature and methods developed in the course of this work, in order to provide a clear line of thoughts to the reader.

In this thesis, the mathematical notation $\mathbf{x}^t$, meaning $\mathbf{x}$ at time t, was frequently used across the sections of this thesis, and should not be mixed up with raising $\mathbf{x}$ to the power t. Consequently, the notation $\mathbf{x}^0$ means $\mathbf{x}^{t=0}$.

Contents

Chapter 1
Introduction

1.1 Computational Modeling

Mathematical models are used by scientists, engineers, economists and others to describe and analyse phenomena and systems they encounter in their respective disciplines. Many mathematical models do not have simple analytic solutions. In these cases numerical methods are used. They are also used in cases where analysed systems are too large or too complex so that approximated solutions are inavoidable.

Computational modeling is using computers and computer programs to calculate numerical solutions of mathematical models. Computational models are called physical when they are based on a mathematical model of physics and they are called non-physical otherwise.

For example, computational modeling is widely used in video games that simulate the physical world as in flight simulators. It is also used to generate visual effects, like water ripples, smoke or explosions. In these applications, the focus is set on the plausibility of the effects and the computation power required to produce them rather than their accuracy. Therefore the models used for these tasks are often non-physical models, fast enough to produce real-time plausible effects on average computers.

Computational modeling is also used in computer animation and movies where processing can be done offline. Since more computational resources can be allocated for the modeling in this case, more accurate models or physical models can be used to generate the effects.

Physical computational models are used in a wide variety of practical areas such as weather forecasting, high energy physics, the design of complex systems such as submarines, mechanical deformation such as in crash tests, and in medicine. General numerical and mathematical problems are encountered in physical modeling such as solving ordinary and partial differential equations, finding the eigenvalues of large matrices and their corresponding eigenvectors and calculating integrals.

In a medical environment, computer models support physicians in diagnosis, monitoring and surgery planning. They are used in the research field to perform virtual experiments (in-silico) prior to real world (in-vivo) experiments, saving time and resources. And they are also used for training of physicians and also for other educational purposes.

Due to the large inter-individual variability and in many cases the intra-individual variability, more emphasis in being put on the development of personalized computational models that can be used in medical applications. Personalized computational modeling adds more challenges to the modeling task for example in defining the parameters needed to describe inter- and intra-individual variability and also to develop methods to measure and quantify these parameters for each individual. Optimization techniques are often used to find model parameters to fit measurement data which can require significant computational power.

For decades the number of transistors on integrated circuits has been doubling every two years in consistency with Moore's law. Even though, the miniaturization of transistors is getting closer to the limits physics allow, multi-core CPU chip designs are compensating the slowing advances in miniaturization.

With this continuous increase of computational power, the complexity and level of details used in computer modeling is increasing and expected to keep increasing in the coming years. The technological advances in the field of micro-electronics has been driving computer modeling towards using more sophisticated and complex models. Nonetheless algorithm complexity remains a key term in developing computer models.

Algorithms complexity refers to the computational resources an algorithm requires. It is often expressed as a function of the size n of its input using the big O notation to describe how well the algorithm scales with more data.

The big O notation provides an upper bound on the growth rate of algorithms. For example the processing time of n data input with an algorithm of $O(n)$ complexity grows linearly with a growing n. But in case of using an algorithm of $O(n^2)$ complexity the processing time grows quadratically with growing n.

In system design, algorithm complexity plays an important role. For instance, the time complexity of an algorithm defines the maximum size of input data n_T that can be processed in a specific time t_T, and alternatively the minimum time needed to process a specific input data size n_N. That puts limits on the computational power needed for the system and thus the practicability and costs of the system.

If two algorithms perform the exact processing but have different complexities, the faster algorithm means less computational power and thus lower costs. It also means more data can be processed with the same costs.

A nice example, is using the fast Fourier transform (FFT) in systems where discrete Fourier transform (DFT) is needed. The discrete Fourier transform (DFT) of n points is of $O(n^2)$ complexity, whereas the equivalent fast Fourier transform (FFT) of the same n points is of $O(n \log n)$ complexity, making it faster and hence the name.

Algorithm complexity will always be a decisive factor in computer modeling, despite that computational power has been increasing since the very first computers at an exponential rate and is expected to keep increasing in the future, since an $O(n)$ algorithm will always perform better than an $O(n^2)$ algorithm for a large enough n.

1.2 Continuum Mechanics

Continuum mechanics is a field of physics that deals with the mechanical analysis of objects modeled as a continuous medium. It allows the description of mechanical behavior and kinematics of a deforming body through differential equations [1]. The elasticity theory, which is a branch of continuum mechanics, is of great importance for this work, because it allows a quantitative description of finite deformations of inhomogeneous, nonlinear and anisotropy materials, such as biological materials, under the influence of stress [2].

In the following sections, some fundamentals of continuum mechanics and the elasticity theory are introduced. For a thorough introduction into continuum mechanics and the elasticity theory, the reader is referred to [3, 4, 5, 1].

1.2.1 Definitions and Physical Laws

Bodies are described in continuum mechanics as an infinite number of particles. Starting from a fully defined reference state at t_0, each of the body particles P can be identified by the reference coordinates $\mathbf{X}^0$ as shown in Figure 1.1 [1]. Because $\mathbf{X}^0$ is bound to a particle of the body material, the components of $\mathbf{X}^0$ are called material coordinates [6]. At time t, also called the current configuration, the trajectory $\mathbf{x}$ of the particle can be given as a function of $\mathbf{X}^0$ and t

$$\mathbf{x} = \chi(\mathbf{X}^0, t) \tag{1.1}$$

and the displacement vector $\mathbf{u}$ is given as (in the Lagrangian description)

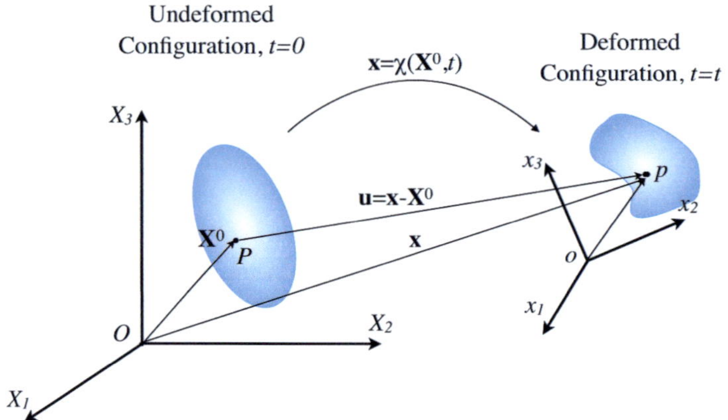

Fig. 1.1 Motion and deformation of a continuum body.

$$\mathbf{u} = \chi(\mathbf{X^0}, t) - \mathbf{X^0} \tag{1.2}$$

The deformation tensor $\mathbf{F}$, is a second-order tensor that quantifies the deformation of a set of adjacent material particles within the body (rotation and deformation) [7, 6]. Consider the line segment $d\mathbf{X} = (dX_1, dX_2, dX_3)$ in the configuration at $t = 0$ shown in Figure 1.2. The deformation gradient tensor $\mathbf{F}$ carries the line segment, $d\mathbf{X}$, into $d\mathbf{x} = \mathbf{F}d\mathbf{X}$. The elements of $\mathbf{F}$ are defined in Eq. (1.3):

$$F_{ij} = \frac{\partial x_i}{\partial X_j} \tag{1.3}$$

The deformation tensor includes a stretch component and a rotation component. Since rotation does not introduce any stress, it should be excluded when deriving a stress-strain relation.

According to the polar decomposition theorem, the deformation tensor $\mathbf{F}$ can be decomposed into an orthogonal rotation tensor $\mathbf{R}$ and a symmetric right stretch tensor $\mathbf{U}$:

$$\mathbf{F} = \mathbf{R}\mathbf{U} \tag{1.4}$$

The right Cauchy Green deformation tensor $\mathbf{C}$ is defined as

$$\mathbf{C} = \mathbf{F}^T\mathbf{F} = (\mathbf{R}\mathbf{U})^T\mathbf{R}\mathbf{U} = \mathbf{U}^T\mathbf{R}^T\mathbf{R}\mathbf{U} == \mathbf{U}^T\mathbf{I}\,\mathbf{U} = \mathbf{U}^T\mathbf{U} \tag{1.5}$$

Since $\mathbf{C}$ is rotation-invariant, it can be used to derive a stress-strain relation that does not include rotation. The Green strain tensor is given with

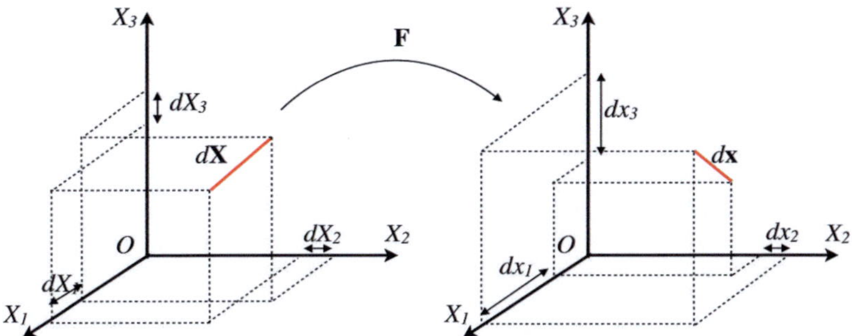

Fig. 1.2 The deformation gradient tensor, $\mathbf{F}$ carries line segment $d\mathbf{X}$ into $d\mathbf{x}$.

$$\mathbf{E} = \frac{1}{2}(\mathbf{F}^{\mathrm{T}}\mathbf{F} - \mathbf{I}) == \frac{1}{2}(\mathbf{U}^{\mathrm{T}}\mathbf{U} - \mathbf{I}) \tag{1.6}$$

$\mathbf{E}$ is used to quantify the strain of a deformed body since it is rotation-invariant and $\mathbf{I}$ is the identity matrix.

Hyperelasticity of a hyperelastic material is described by an energy density function W called the material law or the constitutive law. The Piola-Kirchhoff stress tensor $\mathbf{S}$ can be derived from W and the Green strain tensor $\mathbf{E}$ using

$$S_{\mathrm{ij}} = \frac{\partial W}{\partial E_{\mathrm{ij}}} \tag{1.7}$$

The Piola-Kirchhoff stress tensor relates the forces in the present configuration to areas at the initial configuration. The force $\mathbf{f}$ acting on a surface with area A of a deformed body is given by

$$\mathbf{f} = A^0 \mathbf{F} \mathbf{S}^{\mathrm{T}} \mathbf{n}^0 \tag{1.8}$$

where A^0 is the area of the surface and $\mathbf{n}^0$ is the normal vector on the surface of the initial configuration [1].

The Cauchy stress tensor σ is related to $\mathbf{S}$ according to

$$\sigma = \frac{1}{\det(\mathbf{F})} \mathbf{F} \mathbf{S} \mathbf{F}^{\mathrm{T}} \tag{1.9}$$

The Cauchy stress tensor σ relates the forces of the present configuration to the areas of the present configuration according to

$$\mathbf{f} = A\sigma\mathbf{n} \tag{1.10}$$

where $\mathbf{n}$ is the normal on the surface.

1.3 Modeling Elastomechanics of Soft Tissues

Physically based modeling of deformable objects is an interdisciplinary field of research that combines Newtonian dynamics, continuum mechanics, numerical computation, geometry, vector calculus and computer graphics among many other.

Despite that the field is relatively young, many different physical models which are based on different modeling approaches are already available. These methods can be categorized in:

- Lagrangian mesh-based methods
- Lagrangian mesh free methods
- Reduced deformation models and modal analysis
- Eulerian and semi-Lagrangian methods

The Lagrangian mesh-based methods include models based on continuum mechanics like the method of finite differences, the finite volume method, the boundary element method (BEM), and the finite element method (FEM) which is one of the most popular methods computationally to solve partial differential equations on irregular grids. This category includes also mass-spring systems, relevant to this work described with details in Chapter 2.

For more about physically based deformable models, the reader is referred to the survey of Gibson *et al.* [8] and the survey of Nealen *et al.* [9].

1.4 Medical Imaging

Medical imaging is a vast field of medical engineering encompassing the techniques and processes used to generate images of the human body for clinical or research purposes.

Medical imaging systems or modalities are categorized mainly according to methods used for image acquisition.

The main methods are ultrasonic imaging (US), magnetic resonance imaging (MRI), X-ray computed tomography (CT) and Nuclear medicine imaging.

Due to advances in medical imaging technology, it is possible to routinely acquire high-resolution, three-dimensional (3D) digital images of human anatomy and function using various imaging modalities. Medical image analysis has therefore become a highly active research field [10].

Digital image processing techniques are applied to the acquired images in order to enhance them, extract further information, and to perform segmentation and classification of the objects depicted in the images [11].

1.4.1 Digital Images

An image is a mapping which assigns every spatial point x belonging to a set $\Omega \subset \mathbb{R}^d$ a gray value $b(x)$, where $d \in \mathbb{N}$ denotes the dimension of spatial domain to which the image belongs [12].

Definition 1.1. Let $d \in \mathbb{N}$. A function $b : \mathbb{R}^d \to \mathbb{R}$ is called a d-dimensional image, if

1. b is compactly supported
2. $0 \leq b(x) < \infty$ for all $x \in \mathbb{R}^d$,
3. $\int_{\mathbb{R}^d} b(x)^k dx$ is finite for $k > 0$.

The set of all images is denoted by $\mathrm{Img}(d) := \{b : \mathbb{R}^d \to \mathbb{R} \mid b$ is d-dimensional image$\}$ [12].

To formulate the mathematical definition of digital images, the grids definition must be first introduced.

Definition 1.2. Let $d \in \mathbb{N}$, $\Omega =]0, 1[^d$ and $n_1, \ldots, n_d \in \mathbb{N}$ be some given number. The points $x_{j1,\ldots,jd} = (x_{j1}, \ldots, x_{jd})^{\mathrm{T}} \in \Omega \cup \partial\Omega$ where $1 \leq j_l \leq n_l$ and $1 \leq l \leq d$ and $\partial\Omega$ denote the boundary of Ω, are called grid points. The array $X = (x_{j1,\ldots,jd}) \in \mathbb{R}^{n_1 \times \cdots \times n_d}$ is called the grid matrix [12].

Using definitions 1.1 and 1.2, digital images can be defined with

Definition 1.3. Let $d \in \mathbb{N}$, $b \in \mathrm{Img}(d)$, and let Ω_d be a $n_1 \times \cdots \times n_d$ grid (see Def. 1.2), the arrays $B^\bullet = b^\bullet_{j1,\ldots,jd} \in \mathbb{R}^{n_1 \times \cdots \times n_d}$ where $j_l = 1, \ldots, n_l$ and $l = 1, \ldots, d$ and $b^\bullet$ is the discretization of the continuous image b on the discrete grid Ω_d are called d-dimensional digital images [12].

Two methods are common for the discretization of b on the discrete grid Ω_d: The mesh-point model where $b^\bullet$ is given for every point j with

$$b^\bullet_j = b^\bullet_{j1,\ldots,jd} = b(x_j) \quad \text{for all} \quad x_j \in \Omega_d \tag{1.11}$$

and the mid-point model with

$$b^\bullet_j = b^\bullet_{j1,\ldots,jd} = \int_{c_j} b(x)dx \quad \text{for all} \quad j = 1, \ldots, N. \tag{1.12}$$

where c_j is a small region with center x_j [12].

1.4.2 Ultrasonic Imaging

Ultrasonic imaging is noninvasive, inexpensive, portable and has an excellent temporal resolution. It uses the partial reflection of acoustic waves at interfaces between different tissues as a function of time and velocity of the wave in the medium to generate the image. Other phenomena such as attenuation, dispersion, refraction and scattering occur in ultrasonic imaging and must be taken into account when processing the reflected waves. Doppler principle is also used in Doppler imaging to study not only the morphology or anatomy of tissues but also the function as in blood flow and myocardial velocities [10].

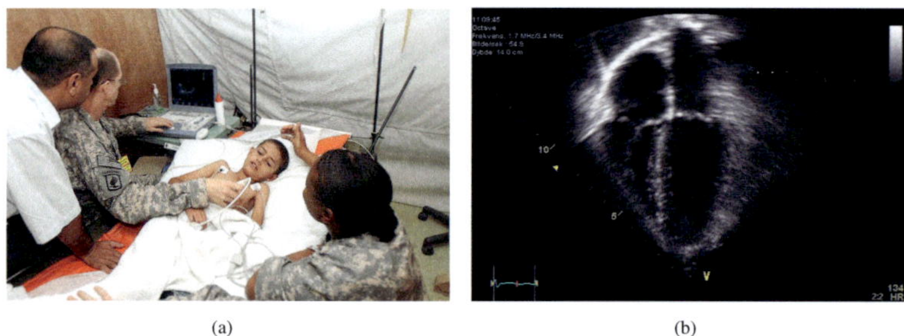

(a)   (b)

Fig. 1.3 Ultrasonic Imaging system [13] (a). An Echocardiography of the heart in four chambers view [14] (b).

1.4.3 Projection X-Ray

Wilhelm Röntgen discovered X-rays and it's abiligy to pass through different sort of materials including biomaterials in 1895. And soon enough it found application in medicine when an X-ray picture of a hand was made.

Projection X-rays are used to detect skeletal system pathologies, as well as in disease processes in soft tissue such as lungs cancer. Projection X-rays are also routinely used for imaging of teeth in dental radiology. Mammography, which is an X-ray examination of the breast, is also routinely used for screening women for breast cancer. For more on the physics of X-ray and the design of X-ray machines the reader is referred to [10, 15, 16].

1.4.4 X-Ray Computed Tomography

X-Ray computed tomography is an imaging modality that uses X-ray attenuation properties of the different tissues in the body to produce cross-sectional images [10]. A thin X-ray beam is used to scan a set of lines of the body, the X-ray beam source rotates around the object and the scanning procedure is repeated for a large number of angles [15, 10]. The acquired data from the scans are then digitally processed to reconstruct the cross-sectional image. CT can also generate three-dimensional images by scanning a number of consecutive slices of the body [10].

1.4.5 Magnetic Resonance Imaging

MRI uses the property of nuclear magnetic resonance (NMR) to image nuclei of atoms inside the body [10].

The object is placed under the effect of a powerful uniform magnetic field $\mathbf{B_0}$ defined as the z-axis of the coordinate system $\mathbf{B_0} = (0, 0, B_0)$, to align the magnetic moments $\boldsymbol{\mu}$ of atoms in the body along the z-axis, mainly hydrogen protons ($_1^1\text{H}$) in water molecules that makes much of the body weight. Each volume element (voxel) of the imaged object contains a huge amount of atoms each having its own spin and the associated magnetic momentum $\boldsymbol{\mu}$. Under the effect of B_0 it can be shown that the behavior of a large amount of spins is equivalent to that of a net magnetization vector $\mathbf{M_0} = (0, 0, M_z)$ [10]. In each voxel in this state, called the the dynamic equilibrium state, we have

$$\mathbf{M_0} = \sum_{i=1}^{n_s} \boldsymbol{\mu}_{iz} \tag{1.13}$$

where n_s is the number of spins in a voxel.

Using electromagnetic radio frequency (RF) waves at resonance frequency of ($_1^1\text{H}$), the equilibrium is disturbed temporarily as atoms absorbs the RF signals photons. As a result the net magnetization x- and y-axis components deviate from zero. The change of direction of $\mathbf{M_0}$ is called the flip angle α. And the net magnetization in this case is denoted $\mathbf{M_\alpha}$. When the RF field is switched off the system returns gradually to the dynamic equilibrium state. This return is called relaxation. Spin-lattice relaxation is the phenomenon that causes the z-axis component of $\mathbf{M_\alpha}$ to grow from $M_0 \cos \alpha$ to M_0. T_1 is the time constant of the spin-lattice relaxation energy phenomenon:

$$M_z = M_0(1 - e^{-t/T_1}) \tag{1.14}$$

T_2 is the time constant of the spin-spin relaxation phenomenon that causes the x- and y-axis components of the net magnetization vector to disappear due to dephasing of spins:

$$\mathbf{M}_{tr} = \mathbf{M}_{tr}(0)e^{-t/T_2} \qquad (1.15)$$

where $\mathbf{M}_{tr}$ is the transverse component of the $\mathbf{M}$.

The relaxation phenomena result in atoms emitting electromagnetic waves with frequencies depending on $w = \gamma B_0$ where γ is a gyromagnetic ratio. Since these signals are not spatially encoded, no useful image can be reproduced of the signals detected by the RF receivers. Therefore, three orthogonal magnetic gradients ($\mathbf{G} = (G_x, G_y, G_z)$)are used for the localization of voxels in the field of view. These additional magnetic fields make the strength of magnetic field dependent on the position within the imaged object, hence making the frequency of the released photons during relaxation dependent on their position. Since position information are encoded in the frequency domain, the image can be generated by applying the inverse Fourier transformation (IFT) [15, 10].

Different tissues have different T_1 and T_2 time constants and therefore return to the dynamic equilibrium at different rates. Mainly, this difference which can be detected is used to generate the intensities of different voxels and thus the image [10].

MRI can be used to generate 2D- or 3D-images depending on the magnetic fields gradients sequences chosen for the imaging task. Diffusion tensor imaging (DTI) is one variant of the MRI that measures the diffusion of water molecules in biological tissue.

Other variants of the MRI imaging modality exists such as real-time MRI, functional MRI (fMRI), tagged cardiac MRI and magnetic resonance angiography (MRA)[10, 15, 16].

1.4.6 Digital Image Processing and Segmentation

After image acquisition, an image processing step is usually required to facilitate the analysis of the medical image. Digital image processing techniques and filtering are generally used for scaling, translating, rotation or coordinate transformation, and also, for image enhancement, edge detection, noise elimination, averaging as well as alignment of images when stacking 2D-images to create 3D-images.

Segmentation refers to the process of detecting a set of pixels belonging to a specific structure in the 2D-image (voxels in the 3D-image) and associating a tag with these pixels (voxels). Figures 1.5(a) shows a 2D MRI slice. After segmentation, pixels belonging to the same tissue class are tagged and the tags can be used to

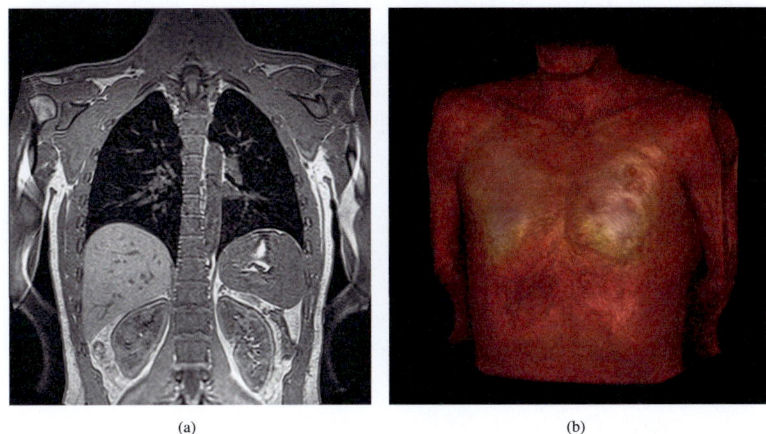

Fig. 1.4 A dataset of the torso of a 27 years old healthy volunteer acquired using a magnetic resonance imaging (MRI) system. A 2D slice is shown in (a) and a 3D reconstruction is shown in (b).

create 3D geometrical models of the anatomy as in Figure 1.5(b).

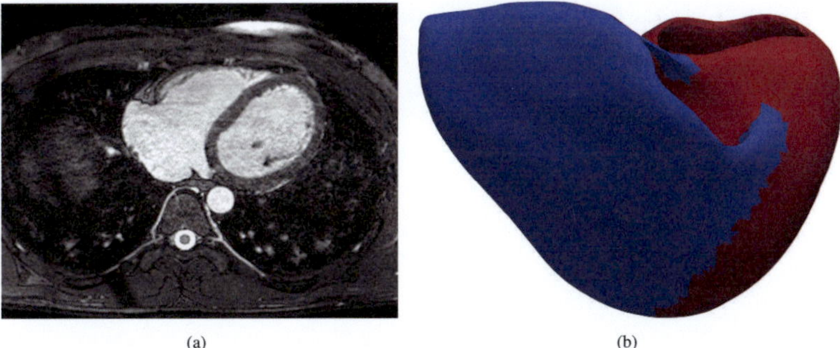

Fig. 1.5 MRI torso dataset showing the heart in a diastolic state with a good contrast between blood and the myocardial tissue (a) allowing for a better segmentation of the heart. The resulting segmented 3D model of left and right ventricles (b).

Manual segmentation is done by experts who assign the tags manually to the pixels (voxels). Therefore, manual segmentation is expensive, time consuming and its accuracy depends strongly on the qualifications and skills of the experts performing the segmentation.

Therefore automated segmentation is of great interest, and has been for many years an active research field that witnessed many successes in the last decade. Although, automated segmentation methods developed explicitly for the segmentation of certain organs or tissue such as the brain, skeletal bones, the ventricles, etc., fully automated hole body segmentation of CT or MRI datasets of humans is still a distant goal.

Chapter 2
Adamss: Advanced Anisotropic Mass-Spring System

2.1 Motivation

Many deformable models based on the solid foundation of continuum mechanics are currently used in a variety of applications that require accurate physically based modeling of deformable objects. The most prominent of these methods is the finite element method (FEM) which is widely used in computational science for solving partial differential equations on irregular grids.

Despite the dominance of the FEM, interest in developing other accurate models still exists. This interest is fueled by the search for real-time methods for producing realistic deformations in computer animations where the high computational costs of FEM make it unsuitable. Interest exists also within other fields of research where accurate but computationally efficient deformable models are needed.

Deformable models are gaining more and more interest in medical environments as models evolve and the computational capacities of hardware increases. For example virtual surgery planning is one of the promising application where deformable models are applied. With virtual surgery planning surgeons may gain the opportunity to test different critical surgery scenarios in a low cost environment without ethical restriction. It may also allow for making predictions about the outcome of a surgery prior to actually performing that surgery, giving the surgeon the opportunity to avoid undesirable outcomes [17, 18, 19, 20, 21]. Modeling soft tissue mechanics is another and important research area in medical engineering where deformable models are used to better understand the behavior of these tissues, especially in cases where experimentation is either unethical or beyond the physical possibility [22, 23, 24, 25, 26].

In many cases, when investigating specific phenomena, a large number of deformation simulations must be conducted. That makes computational efficiency a major factor when choosing a modeling technique. The faster the modeling algorithm is the more simulations can be evaluated within the same timeframe. A tradeoff between accuracy and computational efficiency can be considered in order to rapidly gain a coarse understanding of the general trend of the investigated phenomena. This approach can simplify the further analysis of these phenomena using a more accurate, but also more time consuming modeling methods.

Mass-spring systems are considered the simplest and most intuitive of all deformable models. They are computationally efficient as they require only solving a system of coupled ordinary differential equations. They also handle large deformations and large displacements with ease.

Mass-spring systems are the method of choice for cloth animation [27, 28, 29, 30, 31, 32, 33], they were also used for facial modeling [34, 17], for modeling muscle [35, 36, 37, 38, 39], and in virtual surgery planning [17, 40, 19, 20], and also in segmentation and image registration [41, 42].

In this work, a physically based deformable model is presented. The model is a modified mass-spring system that addresses and solves the problems present in ordinary mass-spring systems.

2.1.1 Mass-Spring Systems

In order to model an object using a mass-spring system, the object is discretized to mass particles p_i $(i = 1, \ldots, n)$, then a network of massless springs connecting the particles together is installed.

Mass-spring systems vary according to the discretization mesh, the way the springs are set between the particles, and the functions used to model the springs.

For example, a simple mass-spring system can be built by discretizing the object using a regular hexahedral mesh, setting the mass particles to the vertices of the hexahedrons and finally setting springs along the sides of the hexahedrons connecting these particles (Fig. 2.1(a)). Another system can be created by adding springs that connect the particles diagonally in each of the hexahedron faces (Fig. 2.1(b)) or diagonally through the volume of the hexahedron connecting particles sitting on opposite vertices in relation to the barycenter (Fig. 2.1(c)).

At any given time t, the state of the system is defined by the positions $\mathbf{x}_i$ and the velocities $\mathbf{v}_i$ of the particles. The force $\mathbf{f}_i$ at a particle p_i is computed according to its spring connections with its neighbors, in addition to external forces such as gravity or friction. The Newton's second law of motion is used to calculate the motion of each particle:

$$m_i \frac{d\mathbf{x}_i}{dt} = \mathbf{f}_i \tag{2.1}$$

and the Newton's second law for the entire particles system can be expressed as

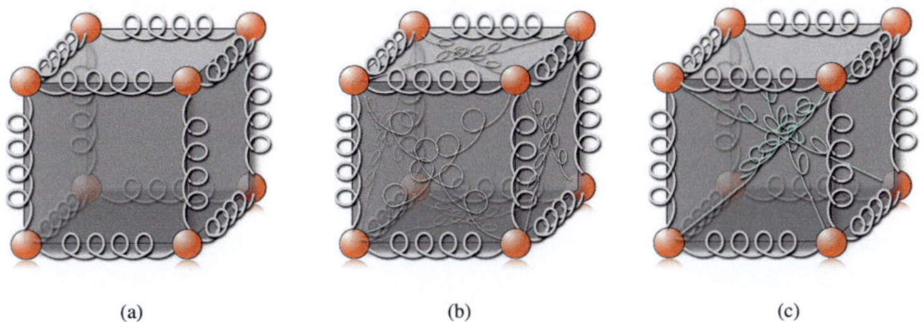

(a)   (b)   (c)

Fig. 2.1 Single hexahedral elements of mass-spring systems with different springs network topologies. Structural springs (a), structural and surface springs (b), structural and diagonal springs (c). Many of these elements are stacked together along a regular three-dimensional mesh to model the geometry and mechanics of the deformable object.

$$\mathbf{M}\frac{d\mathbf{x}}{dt} = \mathbf{f}(\mathbf{x}, \mathbf{v}) \tag{2.2}$$

where $\mathbf{M}$ is a $3n \times 3n$ diagonal matrix also called the mass tensor, $\mathbf{x}$ is a $3n$ vector of the coordinates of all particles.

By solving this system of coupled ordinary differential equations, the coordinates of the particles can be updated as the model deforms.

2.1.2 Problems Associated with Mass-Spring Systems

Ordinary mass-spring systems suffer from several intrinsic limitations that restrict their use in physical modeling. The most troubling of these limitations are listed here:

- In comparison with models based on elasticity theory like the finite element or the finite differences methods, ordinary mass-spring system are not necessarily accurate. Most such systems are not convergent, that is, as the mesh is refined the simulation does not converge on the true solution[9].
- The behavior of these systems depends heavily on the topology and the resolution of the mesh. If the mesh changes the simulation does not converge to the same solution obtained using the original mesh [9].
- Finding the right springs functions and parameters to obtain an accurate model is a very difficult and application dependent process.
- Setting the masses correctly to model homogeneous materials is somewhat troublesome.
- Using ordinary mass-spring systems neither isotropic nor anisotropic materials can be generated and controlled easily.

Concerning the last point, if all springs are set to the same stiffness, anisotropies, which correspond to the mesh geometry chosen for the mass-spring system, will be generated. These anisotropies are generally undesirable.

The anisotropic effect related to the mesh topology decreases with increasing mesh density, if the tiling of the object volume was computed from the triangulation of random uniformly-distributed sample points. However, using an extremely dense mesh contradicts the objective on which using mass-spring systems was based, namely computational efficiency.

It is possible to generate the isotropic or the anisotropic effect in ordinary mass-spring systems by tuning individual spring stiffnesses to reproduce the desired effect [43, 44], or by designing the mesh in order to align springs on some direction of interest [45, 46]. Both methods are time consuming and not applicable to different geometries in a straight-forward manner.

Ordinary mass-spring systems cannot enforce a constant-volume constraint on the modeled object.

Since modeling the deformation of ventricular myocardial tissue is at the heart of this work, and since this tissue exhibits anisotropic mechanical properties along fiber, sheet and sheet-normal directions (this will be discussed in details in Chapter 4) and maintain a constant volume under deformation, it follows that ordinary mass-spring systems are not suitable.

2.1.3 Preliminary Groundwork

Bourguignon *et al.* proposed a method to control anisotropy in mass-spring systems [38] where anisotropies of a deformable object are specified independently from the underlying mesh topology used for mass discretization. In the same paper, Bourguignon proposed a method for volume preservation loosely related to the soft volume preservation constraint of Lee *et al.* [47]. Using this method the volume variations during deformation depend on the parameters used for the materials and on the type of deformation the model undergoes. In applications where these volume variations are considered too high, a hard-constraint on volume preservation must be implemented.

M. Mohr [39] used Bourguignon's method to control anisotropy in combination with an implementation of elasticity theory for every individual voxel to enforce volume preservation and to model the passive mechanical properties of the myocardial tissue. The concept of combining the mass-spring system and the elasticity theory solution using FEM for every voxel has proven to be fruitful. However, the hybrid model of M. Mohr suffered from a major problem which is the further use of mesh springs, including springs along the hexahedral mesh lines, two diagonal springs on each surface and four diagonal springs in each hexahedron. Not only that these springs are incompatible with Bourguignon's method, because they tend to generate unwanted anisotropies, but also, they require new parameters every time the model is scaled or a different mesh density is used, or when a new object is modeled [48]. Furthermore, the use of these springs in combination with the FEM for modeling passive mechanical properties of tissues results in a systematic error in reproducing the passive mechanical properties.

2.1.4 Introducing Adamss

In this work a modified mass-spring system, called Adamss (Advanced anisotropic mass-spring system), based on the work of Bourguignon *et al.* [38] is presented. The system is composed of several building blocks with defined interfaces between the blocks. Figure 2.2 shows the building blocks constituting the system. This design allows for the development of building blocks based on completely

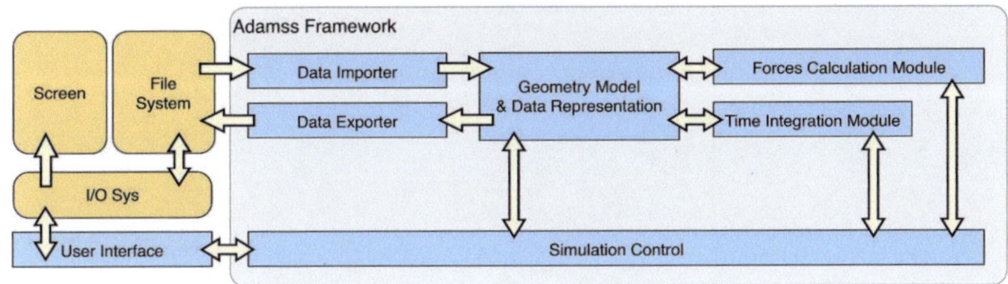

Fig. 2.2 The building blocks of Adamss (Advanced anisotropic mass-spring system), the physically based deformable model presented in this work.

different concepts as long as the interfaces with other blocks are intact making the system flexible and extendible.

During the course of the development of the system, the concept introduced by M. Mohr of combining the theory of elasticity to model the passive mechanical properties of materials and to enforce volume preservation was adopted and extended in the corresponding building blocks. Although the concepts are very similar, the implementations, detailed later in this chapter, differ significantly.

In the following sections the building blocks of the system are presented and the different implementations of each of the building blocks are detailed. Gradually, we will show how the proposed system takes on the mentioned problems associated with mass-spring systems (see Section 2.1.2) and solve them elegantly. Additionally, we will show that the system's behavior is independent of the mesh resolution, and partly independent of the mesh topology. At the end we will show that the overall system is of $O(n)$ complexity.

2.2 Simulation Scheme

Figure 2.3 shows the basic simulation scheme of Adamss.

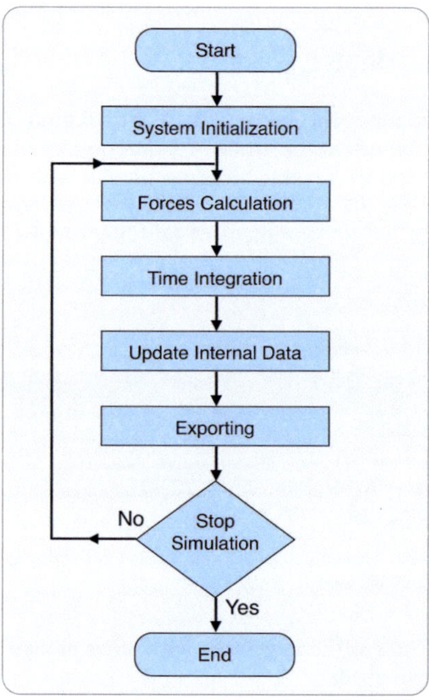

Fig. 2.3 A flowchart of the basic simulation scheme of Adamss.

The simulation start with an initialization step, where the model geometry files including information about anisotropies and boundary conditions, and parameters regulating the mass-spring system generation process are loaded.

After the mass-spring system has been initialized, the system starts the simulation loop. In every iteration of the simulation loop, the different forces working on the particles of the system are calculated. Then time integration of the equations of motion is used to calculate the velocities and offsets of the system's particles.

Depending on the chosen data exporting rate, information about the forces, velocities, and offsets of the system can be exported. Additionally general information like the system's total volume or the system's total kinetic energy can be exported in this step. After that the coordinates of the particles are updated using the offsets and the loop iterates until a condition for stopping the simulation is fulfilled. These conditions can be simulation time crossing a maximum simulation duration t_{max} or kinetic energy dropping below a threshold $E_{\mathrm{k,min}}$. Numerical instability and external interception also stops a running simulation.

The following sections will detail each steps of the simulation scheme presented here.

2.3 Structure Initialization

To initialize the structure of the modified mass-spring system, files containing information about the geometry of the modeled object, and properties of that geometry are imported to the framework.

According to the chosen discretization mesh topology, the modeled object mass is discretized to mass particles as in ordinary mass-spring systems (see Section 2.1.1). Nonetheless, springs are not set on the edge of the discretization mesh.

Instead, the space occupied by the modeled object is also discretized to volume elements according to the selected mesh topology. These resulting volume elements are actually defined by the particles generated during the discretization of the object's mass. Each of the elements encloses a geometrical domain of the object where parameters, important for the modeling, are set and considered homogeneously distributed within that domain. These parameters include the material type of the volume element, the mass-density, different anisotropies, different stiffnesses, the bulk modulus, and other parameters related to the specific material type. The volume elements can take different geometrical shapes depending on the mesh topology used. For example, if a hexahedral mesh topology was used, the resulting volume elements will then be hexahedra. The number of vertices or particles that define a volume element depends on the element type. The faces of a volume element defined by the vertices are called facets.

As soon as defining volume elements is completed, the optional defining of surfaces of enclosed cavities starts. Each of the cavities is defined by the vertices of the facets that enclose the volume of a cavity. These cavities can be used to add constraints to the deformation, for example by enforcing a volume preservation of the cavity volume during deformation. It is also possible to set a constant homogeneous stress to all facets of a cavity, to model ventricular cavities pressure for example.

There are many possibilities to calculate the volume of an enclosed region, providing the surface enclosing the region is known. In this work we used a Delaunay tetrahedralizations technique to generate a tetrahedral mesh of the cavity starting with the vertices marking the cavity. The volume of the cavity can then be calculated by summing the volume of the resulting tetrahedrons.

2.3.1 Mesh Topologies

During the course of development of the modeling framework, the hexahedral mesh topology and several different tetrahedral mesh topologies were implemented. In the following, the implementation of these topologies is presented. The advantages and drawbacks of each of the methods will be discussed later in Section 2.9.

2.3.1.1 Hexahedral Mesh Topology

The hexahedral mesh topology is a regular grid, where after first initialization all mesh elements are identical rectangular hexahedra (Fig. 2.4(b)). In the special case when the length of all hexahedra sides are equal, the mesh elements are called voxels (Fig. 2.4(a)).

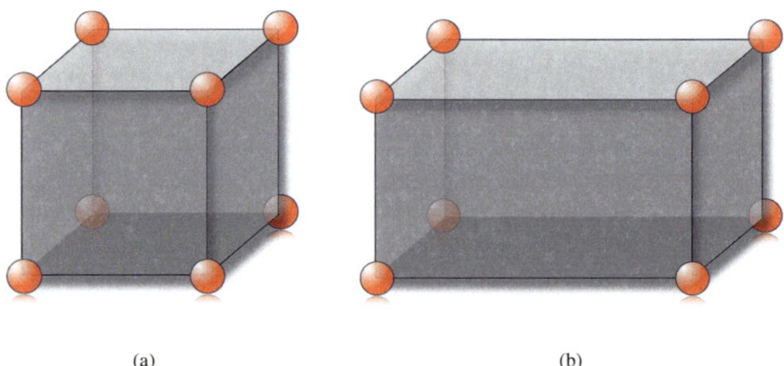

(a)   (b)

Fig. 2.4 Hexahedral mesh topologies: voxels mesh with $1mm$ voxel side(a), rectangular hexahedral mesh of resolution $1mm \times 2mm \times 1mm$ (b).

Here, the mass of the modeled object is discretized according to the mesh into particles that define the vertices of the volume elements of the modeled object. This is not only true for the hexahedral mesh but also for the different tetrahedral meshes.

In order to setup a homogeneous material, the mass of each particle is computed according to the volume of the Voronoi region around it [43]. In the general case it can be calculated with

$$m_i = \sum_{k=1}^{n} \frac{\rho_k V_k}{N_k} \tag{2.3}$$

where m_i is the mass of the particle p_i, n is the number of volume elements neighboring the particle. V_k, ρ_k and N_k are respectively the volume, the mass density, and the number of vertices of the volume element $\mathcal{V}_k$.
For a hexahedral mesh m_i is given by

$$m_i = \frac{V_v}{8} \sum_{k=1}^{n} \rho_k \tag{2.4}$$

where n is the number of hexahedra neighboring p_i and ρ_k is the mass density of the hexahedron $\mathcal{H}_k$. V_v is the volume of a hexahedron of the hexahedral mesh.

To control anisotropy, the method presented by Bourguignon *et al.* [38] is used. The basic idea is to define several axes of mechanical anisotropy in the barycenter of each volume element of the mesh. Forces generated due to the deformation of the model will act only in the direction of these axes. For example, to model the mechanical deformation of a muscle, the direction of anisotropy axes must be set in the direction of the fibers of the muscle in each of the volume elements of the model's mesh. Although the method allows the definition of several axes of anisotropy, only three were defined in each of volume elements. This was enough to model the anisotropic mechanical behavior of all types of materials modeled with the developed framework.

In a volume element, each axis intersects with the faces of that element in two points, called intersection points (Fig. 2.5(a)). During deformation, the axes evolve with the volume elements to which they belong. At any given moment t, the orientation of an axis ζ_l in a volume element $\mathcal{V}_k$ can be determined using the line extended between the pair of intersection points the axis defines. This can be done numerically by calculating the vector between the pair (q_{2l}, q_{2l+1}) regardless of its direction. The vector is given by

$$\zeta_l^t = \mathbf{x}_{2l}^t - \mathbf{x}_{2l+1}^t \tag{2.5}$$

For this reason, intersection points are used to track the axes of anisotropies during the deformation of the model.

To compute the coordinates of the intersection points at any moment t, and hence the orientation of the anisotropy axes, the coordinates $\mathbf{x}_j^t$ of each intersection point q_j are given by a linear interpolation of the coordinates of the face vertices to which it belongs, using the rectangle linear interpolation shape function

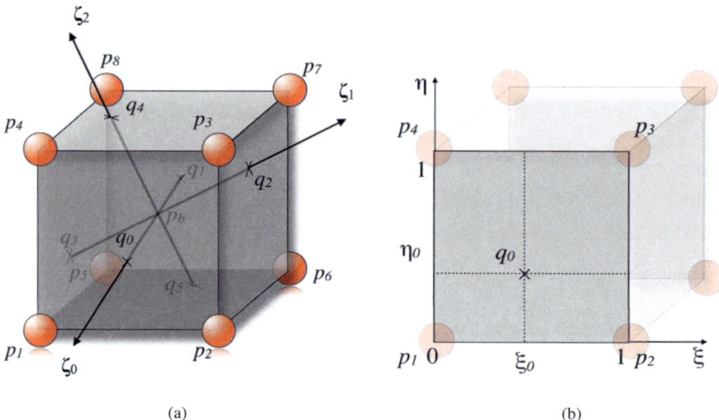

(a)   (b)

Fig. 2.5 Intersection points in a hexahedral volume element: The 3D hexahedral element with three axes of anisotropy set at the barycenter and the six intersection points they define (a), a rectangle face of the element containing the intersection point q_0 and the coefficients ξ_0 and η_0 related to q_0, the vertices of that face are denoted with p_i $(i = 1, \ldots, 4)$ (b).

$$\mathbf{x}_j^t = \sum_{i=1}^{4} N_i(\xi_j, \eta_j)\mathbf{x}_i^t \tag{2.6}$$

where $\mathbf{x}_i^t$ are the coordinates of the vertices of the rectangular face $\mathcal{F}_j$ to which q_j belongs and $N_i(\xi_j, \eta_j)$ are the rectangle linear interpolation shape functions [49, 1] and are given with

$$
\begin{aligned}
N_1(\xi_j, \eta_j) &= (1 - \xi_j)(1 - \eta_j) \\
N_2(\xi_j, \eta_j) &= \xi_j(1 - \eta_j) \\
N_3(\xi_j, \eta_j) &= \xi_j\eta_j \\
N_4(\xi_j, \eta_j) &= (1 - \xi_j)\eta_j
\end{aligned} \tag{2.7}
$$

where ξ_j, η_j are the interpolation coefficients associated with q_j (Fig. 2.5(b)).

During structure initialization, the axes of mechanical anisotropies are defined at the barycenter of each hexahedron of the model. The intersection points and the corresponding shape functions are also computed at this stage using the method described here:

1. For each hexahedron $\mathcal{H}_k$ of the mesh, repeat the following steps:
2. For each of the three axes ζ_l of $\mathcal{H}_k$, repeat the following steps:
3. For each face $\mathcal{F}_j$ of $\mathcal{H}_k$, repeat the following steps to locate intersection points axis ζ_l defines:

4. Compute the point where the line extending from the barycenter p_{bk} of $\mathcal{H}_k$ and which is parallel to ζ_l intersects with the plane containing $\mathcal{F}_j$
5. If the computed point was outside $\mathcal{F}_j$, it is not a valid intersection point, go to step 3 (iterate over faces)
6. If the point is in $\mathcal{F}_j$, it is a valid intersection point, compute the related shape function values
7. If two valid intersection points were found for ζ_l, go to step 2 (iterate over axes)

Step 4 is actually a ray tracing task that requires computing the barycenter, which can be done using

$$\mathbf{x}_b = \frac{1}{8} \sum_{i=1}^{8} \mathbf{x}_i \tag{2.8}$$

where $\mathbf{x}_i$ are the hexahedron's vertices. The vector equation of the ray starting from p_{bk} in the direction of axis ζ_l can be given with

$$\mathbf{v} = d\widehat{\zeta_l} + \mathbf{x}_{bk} \tag{2.9}$$

where $\widehat{\zeta_l}$ is the unit vector in direction of ζ_l and d is a distance along ζ_l. The plane of $\mathcal{F}_j$ can be expressed in vector notation with

$$(\mathbf{v} - \mathbf{x}_i) \cdot \mathbf{n} = 0 \tag{2.10}$$

where $\mathbf{x}_i$ is one of the face's vertices, i. e. a random point in the plane. $\mathbf{n}$ is the normal on the plane and can be simply calculated using the face's vertices with

$$\mathbf{n} = \frac{(\mathbf{x}_1 - \mathbf{x}_2) \times (\mathbf{x}_2 - \mathbf{x}_3)}{\|(\mathbf{x}_1 - \mathbf{x}_2) \times (\mathbf{x}_2 - \mathbf{x}_3)\|} \tag{2.11}$$

By substituting Eq. (2.9) in Eq. (2.10) and solving for d we obtain the distance along ζ_l to the plane starting from p_{bk}:

$$d = \frac{(\mathbf{x}_i - \mathbf{x}_{bk}) \cdot \mathbf{n}}{\zeta_l \cdot \mathbf{n}} \tag{2.12}$$

Now by substituting d in Eq. (2.9) the point where the ray intersects with the plane is obtained.

A special method has been developed, not only to check if a specific point which is coplanar to a rectangle is inside that rectangle (step 5), but also to compute the corresponding shape functions (step 6).

First the surface $S_\square$ of the face defined by vertices p_i $(i = 1, \ldots, 4)$ is calculated with

$$S_\square = \|\mathbf{x}_1 - \mathbf{x}_2\| \cdot \|\mathbf{x}_1 - \mathbf{x}_3\| \tag{2.13}$$

where $\mathbf{x}_i$ are the rectangle vertices' coordinates. Then the surfaces of the four different triangles the intersection point q_j defines with the rectangle's vertices, namely $S_{\triangle j12}$, $S_{\triangle j14}$, $S_{\triangle j23}$ and $S_{\triangle j34}$ are calculated (see Fig. 2.5(b)).
The surface of a triangle can be calculated using

$$S_{\triangle 123} = \frac{1}{2}\|(\mathbf{x}_1 - \mathbf{x}_2) \times (\mathbf{x}_2 - \mathbf{x}_3)\| \tag{2.14}$$

here $\mathbf{x}_i$ $(i = 1, \ldots, 3)$ are the triangle vertices' coordinates. If the statement:

$$S_{\square} = S_{\triangle j12} + S_{\triangle j14} + S_{\triangle j23} + S_{\triangle j34} \tag{2.15}$$

is true, then the point q_j is located inside the rectangle, and the shape functions coefficients ξ_j and η_j can be calculated with

$$\begin{aligned}
\xi &= 2 \cdot S_{\triangle j23}/S_{\square} \\
\eta &= 2 \cdot S_{\triangle j34}/S_{\square}
\end{aligned} \tag{2.16}$$

The shape functions of each of the intersection points of a hexahedron $\mathcal{H}_k$ can be arranged in a matrix $\mathbf{C}_k$, we call the coefficient matrix, according to

$$C_{ij} = \begin{cases} N_{ij} \text{ where } p_i \text{ is a vertex of } \mathcal{H}_k \text{ and also the} \\ \quad \text{face } \mathcal{F}_j \text{ containing intersection point } q_j \\ 0 \text{ otherwise} \end{cases} \tag{2.17}$$

N_{ij} is the shape function associated with q_j and p_i and calculated using Eqs. (2.7) and (2.16). There are six intersection points and eight nodes, that means the coefficients matrix $\mathbf{C}_k$ is an 8×6 matrix. Using $\mathbf{C}_k$, it is possible to calculate the coordinates $\mathbf{x}_i^t$ of intersection points p_j^t at time t by calculating

$$\mathbf{x}_j^t = \sum_{i=1}^{8} C_{ij}\mathbf{x}_i^t \tag{2.18}$$

2.3.1.2 Tetrahedral Mesh Topologies

Tetrahedral mesh topologies are topologies where the mesh elements are tetrahedra.

As in hexahedral mesh topologies, the mass of each particle is computed according to the volume of the Voronoi region around it according to Eq. (2.3), which can be rewritten for a the special case of a tetrahedral mesh as

$$m_i = \frac{1}{4}\sum_{k=1}^{n} \rho_k V_k \tag{2.19}$$

where m_i is the mass of the particle i, n is the number of tetrahedra neighboring the particle i, ρ_k and V_k are the mass density and the volume of the tetrahedron $\mathcal{T}_k$.

Similarly to the method used to control mechanical anisotropies in objects modeled using a hexahedral mesh, three axes of anisotropy are defined at the barycenter of each tetrahedron. Each axis defines two intersection points on the triangle faces of the tetrahedron (Fig. 2.6(a)).

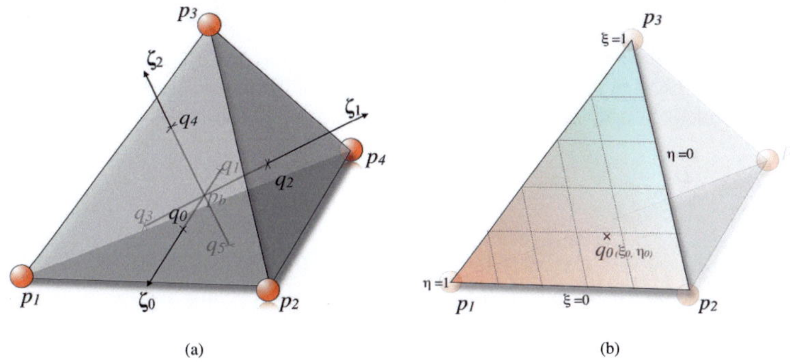

Fig. 2.6 Intersection points in a tetrahedral volume element: The tetrahedron with three axes of anisotropy set at the barycenter and the six intersection points that they define (a), a triangular face of the element containing an intersection point and the coefficients ξ_0 and η_0 related to the intersection point. Note that ξ increases with the cyan color gradient starting from $\xi = 0$ at the line segment (p_1, p_2) and is equal to $\xi = 1$ at p_3, while η increases along the orange color gradient starting from $\eta = 0$ at (p_2, p_3) until it reaches $\eta = 1$ at p_1 (b).

To track the axes as the model and the underlying tetrahedra deform, the coordinates of the intersection points are calculated as a linear interpolation of the vertices coordinates of the triangle faces to which the intersection points belong with

$$\mathbf{x}_j^t = \sum_{i=1}^{3} N_i(\xi_j, \eta_j)\mathbf{x}_i^t \qquad (2.20)$$

Here, the triangle linear interpolation shape functions $N_i(\xi_j, \eta_j)$ are used [1]:

$$\begin{aligned} N_1(\xi_j, \eta_j) &= 1 - \xi_j - \eta_j \\ N_2(\xi_j, \eta_j) &= \xi_j \\ N_3(\xi_j, \eta_j) &= \eta_j \end{aligned} \qquad (2.21)$$

where $\mathbf{x}_j^t$ is the coordinate of the intersection point q_j at a time t, and ξ_j, η_j are the interpolation coefficients associated with q_j (Fig. 2.6(b)).

During structure initialization, the axes of mechanical anisotropies are defined at the barycenter of each tetrahedron of the model. The intersection points and the corresponding shape functions are also computed at this stage using the same method described in Section 2.3.1.1 for finding intersection points and calculating shape functions, however, after adapting the method to support tetrahedral meshes.

Here also a ray tracing task is performed starting from the barycenter of a tetrahedron in the direction of the anisotropy axes in order to search for intersection points.

The first difference between the hexahedral and the tetrahedral mesh implementation of the method is calculating the barycenter of the mesh element which must be calculated for tetrahedral elements with

$$\mathbf{x}_b = \frac{1}{4} \sum_{i=1}^{4} \mathbf{x}_i \tag{2.22}$$

To check if a specific traced point q_j which is coplanar to one of the triangular faces of the tetrahedron $\mathcal{T}_k$ is inside that triangle and to compute the corresponding shape functions at the same time, the surfaces of triangles $S_{\triangle 123}$, $S_{\triangle j12}$, $S_{\triangle j13}$ and $S_{\triangle j23}$ are calculated using Eq. (2.14). The triangle $S_{\triangle 123}$ is defined by the vertices p_i $(i = 1, \ldots, 3)$, while $S_{\triangle j12}$, $S_{\triangle j13}$ and $S_{\triangle j23}$ are the triangles which the point q_j defines with the face vertices.

If the statement:
$$S_{\triangle 123} = S_{\triangle j12} + S_{\triangle j13} + S_{\triangle j23} \tag{2.23}$$
was true, then the point q_j is located inside the triangular face, and the shape functions coefficients ξ_j and η_j can be given by

$$\xi = S_{\triangle j13}/S_{\triangle 123} \tag{2.24}$$
$$\eta = S_{\triangle j12}/S_{\triangle 123}$$

Setting the shape functions using Eq. (2.24) is legitimate only for linear tetrahedra i.e. straight-sided triangles. The coefficients in this case are also called area or areal coordinates. Eq. (2.24) does not carry over to general isoparameteric higher order triangles i.e. with curved sides.

Here also, the shape functions of each of the intersection points of a tetrahedron $\mathcal{T}_k$ can be arranged in a coefficient matrix $\mathbf{C}_k$, according to Eq. (2.17) where N_{ij} is the shape function associated with intersection point q_j and the vertex p_i and calculated using Eqs. (2.21) and (2.24). In each tetrahedron, there are six intersection points and four nodes, that means the coefficients matrix $\mathbf{C}_k$ is a 4×6 matrix.

Using the matrix $\mathbf{C}_k$, it is possible to calculate the coordinates of intersection points q_j^t at time t by calculating

$$\mathbf{x}_j^t = \sum_{i=1}^{4} C_{ij} \mathbf{x}_i^t \qquad (2.25)$$

Tetrahedral topologies can be generated by systematically dividing a regular grid of the modeled object to tetrahedrons or by using unstructured grids where the modeled object is broken down to irregular tetrahedra using a mesh generation algorithm.

2.3.1.3 Tetrahedral Mesh Topologies Based on Regular Grids

This method take the advantages that regular grids offer, especially the analogy to medical images data structures, and at the same time benefit of the use of tetrahedra to solve problems associated with hexahedra. Each hexahedron of the hexahedral mesh is divided to a number of non-intersecting linear tetrahedra according to a specific division scheme. This strategy was inspired by a method to compute the volume of a deformed hexahedron by dividing it to six non-intersecting tetrahedra and then summing up the computed volume of each of the tetrahedra [50].

Schemes to divide each hexahedron to five or six non-intersecting tetrahedra were implemented. These schemes do not alter the number of vertices of the model nor the masses the hexahedral mass discretization defines.

Figure 2.7 shows a hexahedron divided to five tetrahedra. It is notable that tetrahedron $\mathcal{T}_5$ has a bigger volume in comparison with the remaining tetrahedra.

Figure 2.8 shows a hexahedron divided to six tetrahedra. Using this division scheme all resulting tetrahedra have the same volume.

Figures 2.9(a) and 2.9(b) shows two different mesh topologies based on five tetrahedra per hexahedron schemes. In the the first scheme 2.9(a) all hexahedra were divided identically, while in the second 2.9(b) a division scheme that has a cubic pattern kernel of $2 \times 2 \times 2$ hexahedra is used. In each of the kernel the division scheme produces mirrored tetrahedra.

Figures 2.10(a) and 2.10(b) show two different mesh topologies based on the six tetrahedra per hexahedron scheme. In the the first scheme (Fig. 2.10(a)) all hexahedra were divided identically, while in the second (Fig. 2.10(b)) a division scheme that has a cubic pattern kernel of $2 \times 2 \times 2$ hexahedra is used. In each of the kernel the division scheme produces mirrored tetrahedra.

The schemes presented in Figures 2.9(b) and 2.10(b) were implemented to eliminate anisotropies resulting from the implementation of volume preservation forces that consider a homogeneous elements volume distribution.

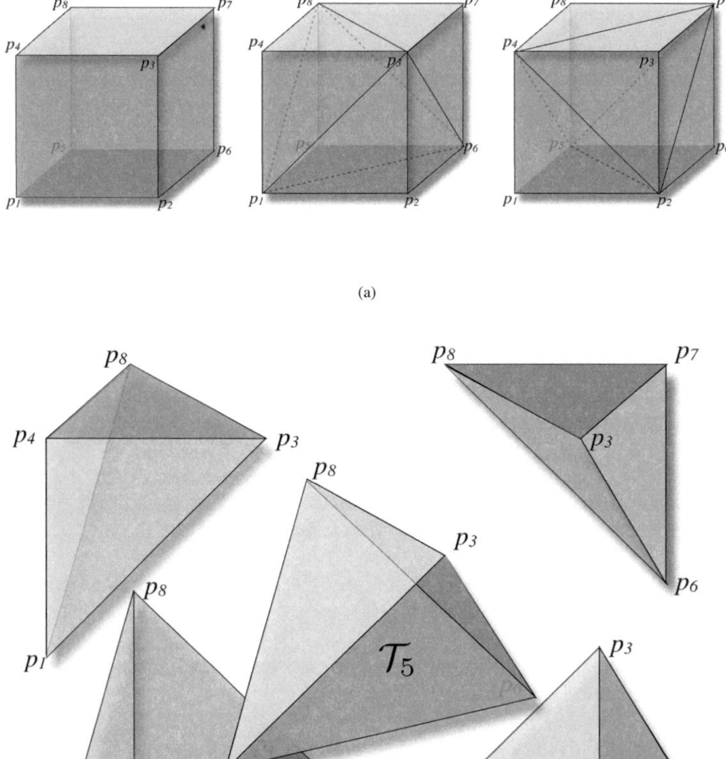

(a)

(b)

Fig. 2.7 In (a), A hexahedron (left) divided to five tetrahedra in two different ways (middle and right). The lines connecting the vertices of the hexahedra define the different tetrahedra resulting from the division scheme. In (b), the tetrahedra resulting from the first division scheme (a, middle) are shown.

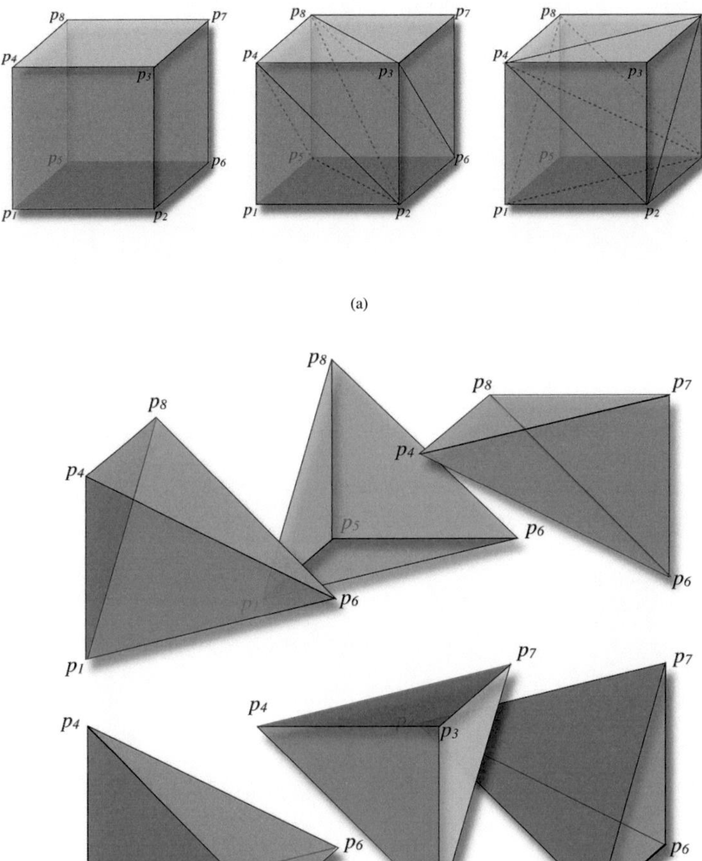

(a)

(b)

Fig. 2.8 In (a), A hexahedron (left) divided to six tetrahedra in two different ways (middle and right), two possible way are not depicted here. The lines connecting the vertices of the hexahedra define the different tetrahedra resulting from the division scheme. In (b), the tetrahedra resulting from the second division scheme (a, right) are shown.

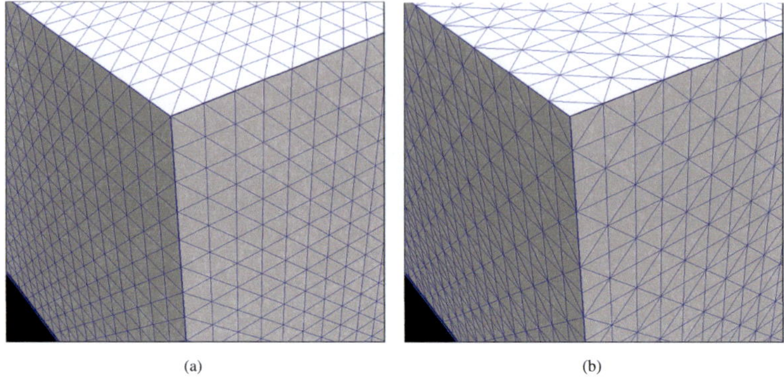

(a)   (b)

Fig. 2.9 Different mesh topologies based on the hexahedron to five tetrahedra division scheme with a model of a cube of $18 \times 18 \times 18 = 5832$ voxels thus to 29160 tetrahedra. All hexahedrons are divided identically (a). A division scheme with a cubic pattern kernel generates a symmetric mesh topology (b).

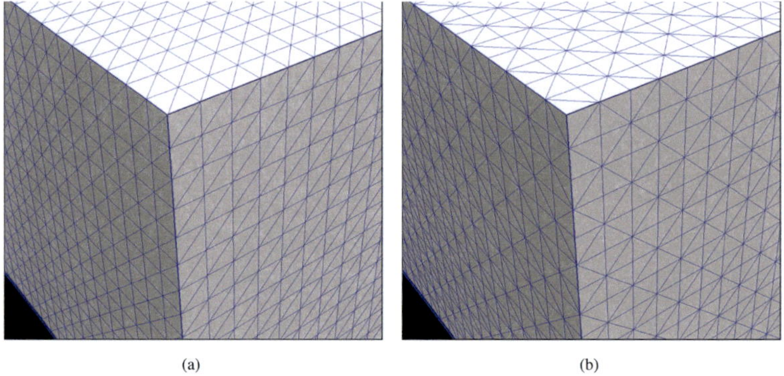

(a)   (b)

Fig. 2.10 Different mesh topologies based on the hexahedron to six tetrahedra division scheme with a model of a cube of $18 \times 18 \times 18 = 5832$ voxels thus to 34992 tetrahedra. All hexahedrons are divided identically (a). A division scheme with a cubic pattern kernel generate a symmetric mesh topology (b).

2.3.1.4 Tetrahedral Mesh Based on Unstructured Grids

Unstructured grids are widely used in computational modeling specially when the modeled object has an irregular shape. In this work, CGAL (Computer Geometry Algorithm Library) [51] was used to generate the tetrahedral meshes out of the lattice image datasets. CGAL is an open source project that provides access to many geometric algorithms in form of a C++ library which is a popular library that is used in many projects [52, 53, 54].

2.4 Forces Calculation

In this section, the different models of forces that act on the particles of the modeled object are discussed. In our implementation forces are divided to internal and external forces.

Internal forces are generated in the volume elements of the modeled object either in response to deformation, or as the result of internal processes active within the element, like the contraction forces of a muscle for example. External forces are forces affecting the particles regardless of the state of the elements to which these particles belong. This group includes forces resulting from external acceleration, initial velocity, external pressures or stresses on the defined cavities of the model. Friction is also part of this group.

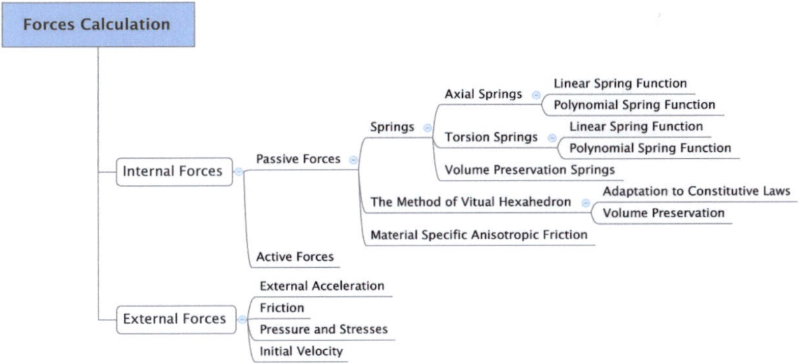

Fig. 2.11 The organization of the force calculating module, showing the models used for the calculation of internal and external forces

Usually, modeling complex objects composed of several materials is of high interest. Modeling cardiac mechanics is one example related to this work. These materials exhibit a variety of mechanical properties. Therefore different forces models must be used in combination to reproduce the object's complex mechanical properties. That also applies for the different internal processes that translate into forces. Objects are often under the effect of several external forces acting at the same time. A combination of different forces models must also be used to model these external forces. Interchangeable internal and external forces models were developed. Each of the models has its area of application, advantages and surely disadvantages. Figure 2.11 shows the organization of these different models.

Before we go through the mentioned models, the method used to control anisotropy in this framework must be explained.

2.4.1 Controlling Anisotropy

In ordinary mass-spring systems, the length of each structural spring set between two particles changes when the model deforms. This results in forces acting on the particles in the direction opposite to the change, trying to bring the spring to its initial length, thus, bringing the system back to equilibrium. However these forces give rise to undesirable anisotropies working in the direction of mesh lines, as mentioned in Section 2.3.

In this work, the method proposed by Bourguignon *et al.* [38] to control anisotropy in mass-spring systems is used. This method provides the possibility to let forces act along pre-defined axes of interest, i.e. along pre-defined axes of anisotropy.

As described in details in Section 2.3, in each volume element $\mathcal{V}_k$ of the model, three axes of anisotropy ζ_l $(l = 1, \ldots, 3)$ are defined at the barycenter. These axes intersect with the surfaces of the volume element in intersection points. For a volume element $\mathcal{V}_k$, the coordinate $\mathbf{x}_j$ of an intersection point q_j can be given using the coefficient matrix $\mathbf{C}_k$ of that element using Eq. (2.18) or Eq. (2.25), according to the mesh topology. For both the hexahedral and the tetrahedral, we can write

$$\mathbf{x}_j^t = \sum_{i=1}^{n} C_{ij} \mathbf{x}_i^t \tag{2.26}$$

where n is the number of vertices of $\mathcal{V}_k$. Additionally, it is possible to calculate the velocity of intersection points with

$$\frac{d\mathbf{x}_j^t}{dt} = \sum_{i=1}^{n} C_{ij} \frac{d\mathbf{x}_i^t}{dt} \tag{2.27}$$

To control anisotropy, forces are first calculated at intersection points and then distributed to the particles. For instance, the force $\mathbf{f}_j$ calculated at intersection point q_j is distributed to the particles p_i of the face to which q_j belongs according to the shape functions $N_i(\xi_j, \eta_j)$ calculated for q_j using Eq. (2.6) or Eq. (2.20) depending on the mesh topology. The portion of $\mathbf{f}_j$ acting on vertex p_i is given with

$$\mathbf{f}_{ij} = N_i(\xi_j, \eta_j)\, \mathbf{f}_j \tag{2.28}$$

The force $\mathbf{f}_i$ active at vertex p_i of a volume element $\mathcal{V}_k$ is the accumulation of portions of the forces $\mathbf{f}_{ij}$ acting on intersection points belonging to faces of which p_i is a vertex. By making use of the coefficient matrix $\mathbf{C}_k$, and since we have six intersection points per volume element, the last statement can be formulated mathematically as

$$\mathbf{f}_i^t = \sum_{j=0}^{5} C_{ij} \mathbf{f}_j^t \tag{2.29}$$

To model an anisotropic behavior, stresses σ_j acting on faces of $\mathcal{V}_k$ should be transformed to forces $\mathbf{f}_j$ at the intersection points q_j of the faces, and then distributed using 2.29 to the particles p_i of $\mathcal{V}_k$. To model isotropic behavior, stresses σ_j acting on faces of $\mathcal{V}_k$ should be transformed to forces acting directly on particles p_i of $\mathcal{V}_k$ ignoring the axes of anisotropies, intersection points and forces distribution rule.

2.4.2 Internal Forces

Internal forces can either refer to forces generated in the volume elements of the model due deformation (*deformation forces*), or forces generated due to internal processes in the volume elements of the model (*active forces*).

2.4.2.1 Axial and Torsion Springs

One method to calculate deformation forces is using *axial* and *torsion* springs as presented in the work of Bourguignon *et al.*[38].

In each volume element $\mathcal{V}_k$, three axial springs $\mathcal{S}_l$ are defined between each pair of intersection points (q_{2l}, q_{2l+1}). A torsion spring τ_{lm} is defined between each pair of these axial springs $(\mathcal{S}_l, \mathcal{S}_m)$. In total, three axial springs and three torsion springs are defined in each volume element that uses this method to calculate internal forces. Figure 2.12 shows a tetrahedron with three axial springs and three torsion springs. The initial length of an axial spring can be computed using Eq. (2.5) with $t = 0$ and then calculating the Euclidean norm

$$l_{\mathcal{S}_l}^0 = \|\boldsymbol{\zeta}_l^0\| = \|\mathbf{x}_{2l}^0 - \mathbf{x}_{2l+1}^0\| \tag{2.30}$$

where $\mathbf{x}_{2l}^0$ and $\mathbf{x}_{2l+1}^0$ are the initial coordinates of the intersection points defined by the anisotropy axis ζ_l.

By defining the unit vector $\hat{\zeta}_l^t$ in the direction of axis ζ_l^t at time t with

$$\hat{\zeta}_l^t = \frac{\zeta_l^t}{\|\zeta_l^t\|} \tag{2.31}$$

the angle α_{lm}^t between the axes ζ_l and ζ_m can be given by

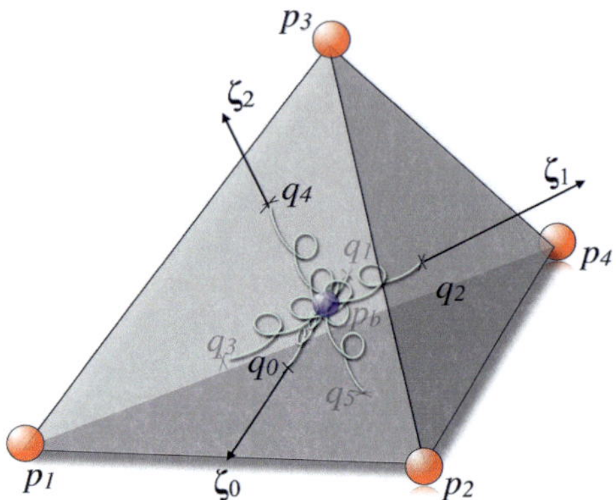

Fig. 2.12 A tetrahedron with three axial springs (in cyan) along the axes of anisotropy and three torsion springs in the barycenter of the tetrahedron (in violet).

$$\alpha_{\mathrm{lm}}^t = \arccos\left(\hat{\boldsymbol{\zeta}}_{\mathrm{l}}^t \cdot \hat{\boldsymbol{\zeta}}_{\mathrm{m}}^t\right) \tag{2.32}$$

and the initial angle α_{lm}^0 is computed by setting $t = 0$ in Eq. (2.32).

Forces $\mathbf{f}_{2l}$ and $\mathbf{f}_{2l+1}$ the springs exert on intersection points q_{2l} and q_{2l+1} defined by ζ_l are composed of an axial component $\mathbf{f}_S$ and two torsion components $\mathbf{f}_\tau$ each. A general formulation for these forces can be given by

$$\mathbf{f}_{2l} = \mathbf{f}_S(\boldsymbol{\zeta}_l, \alpha_{\mathrm{lm}}, \alpha_{\mathrm{ln}})\hat{\boldsymbol{\zeta}}_l + \mathbf{f}_\tau(\boldsymbol{\zeta}_l, \alpha_{\mathrm{lm}}, \alpha_{\mathrm{ln}})\hat{\boldsymbol{\zeta}}_m + \mathbf{f}_\tau(\boldsymbol{\zeta}_l, \alpha_{\mathrm{lm}}, \alpha_{\mathrm{ln}})\hat{\boldsymbol{\zeta}}_n \tag{2.33}$$

$$\mathbf{f}_{2l+1} = -\mathbf{f}_{2l} \tag{2.34}$$

with $l \neq m \neq n$.

These components vary depending on spring functions used. Different spring functions can be used to reflect the mechanical properties of the modeled materials. In the matter of fact, spring functions can be specifically designed to obtain a better fit to mechanical properties of the modeled material.

Linear Axial Springs

Linear springs are commonly used in ordinary mass-spring systems. They are simply an implementation of Hooke's law where force is a linear function of the spring's length:

$$\mathbf{f}_{2l}^t = -k_l(\|\zeta_l^t\| - \|\zeta_l^0\|)\hat{\zeta}_l^t \tag{2.35}$$

where k_l is the stiffness constant of the linear spring.

Higher-Order Axial Springs

Most materials exhibit a non-linear strain-stress relationship, and can be considered linear only for short strain ranges. For instance, the relation can follow a quadratic or a logarithmic curve. Spring functions can be designed depending on the properties of the modeled material by means of curve fitting. By fitting a polynomial function to the strain-stress curve, a polynomial spring function would have the following form:

$$\mathbf{f}_{21}^t = - \left(\sum_{i=1}^{n} k_{l,i} (\|\zeta_l^t\| - \|\zeta_l^0\|)^i \right) \hat{\zeta}_l^t \qquad (2.36)$$

where n is the rank of the function and the parameters k_i are the different stiffness parameters of the polynomial function. From Eq. (2.36) one can easily conclude that more parameters have to be found for spring functions of high order n, complicating the parameterization task. In this work only quadratic and cubic spring functions were implemented.

The parameters k_i of the spring functions can be obtained by fitting the spring function to the strain-stress curves after converting the spring force to stress using a unit surface. But the resulting parameters in this case would be correct only if faces had a unit surface which is clearly not the case. Therefore, the parameters must be scaled using the surface of faces on which the springs exert their forces. The scaled stiffness parameters must be then used in the spring functions mentioned above.

That means, from the software engineering point of view, that memory has to be allocated for the scaled stiffness parameters in each volume element of the model. Additionally, scaling the parameters at the structure initialization phase makes the forces calculation module dependent on the mesh topology module (see Fig. 2.2) which is undesirable because it reduces the flexibility of the framework. By incorporating the surface of the face on which the spring force applies in the forces calculation step, it is possible to avoid the mentioned issues on the cost of increasing the computation time.

The unscaled stiffness parameters E_l found by fitting the spring function to the strain-stress relation have a unit of pressure $[N/m^2]$. Using these parameters, Eq. (2.35) can be rewritten as

$$\mathbf{f}_{2l}^{t} = -\, k_{l}^{t}(\|\zeta_{l}^{t}\| - \|\zeta_{l}^{0}\|)\hat{\zeta}_{l}^{t} \tag{2.37}$$

$$k_{l}^{t} = E_{l}\, A_{j}^{t}\, |\mathbf{n}_{j}^{t} \cdot \hat{\zeta}_{l}^{t}| \tag{2.38}$$

where A_{j}^{t} is the surface of the face $\mathcal{F}_{j}$ on which the force acts and $\mathbf{n}_{j}^{t}$ is the normal on $\mathcal{F}_{j}$ all calculated at time t. Using Eq. (2.37) the notion $\mathbf{f}_{2l} = -\mathbf{f}_{2l+1}$ of Eq. (2.34) becomes invalid, because, for this case, k_{l}^{t} depends on the area and normal of the face $\mathcal{F}_{j}$. Therefore, the spring forces must be calculated independently for the intersection point q_{2l} and q_{2l+1} defined by axis ζ_{l}. The same can be applied to Eq. (2.36) by substituting k_{l} with k_{l}^{t} given in Eq. (2.38). And the resulting polynomial spring function can be given with

$$\mathbf{f}_{2l}^{t} = -\left(\sum_{i=1}^{n} k_{l,i}^{t}(\|\zeta_{l}^{t}\| - \|\zeta_{l}^{0}\|)^{i} \right)\hat{\zeta}_{l}^{t} \tag{2.39}$$

Linear Torsion Springs

Torsion springs act between two pairs of intersection points as demonstrated in figure 2.12.

The linear torsion spring function is given by

$$\mathbf{f}_{2l}^{t} = -k_{lm}(\alpha_{lm}^{t} - \alpha_{lm}^{0})\hat{\zeta}_{m}^{t} \tag{2.40}$$

$$\mathbf{f}_{2m}^{t} = -k_{lm}(\alpha_{lm}^{t} - \alpha_{lm}^{0})\hat{\zeta}_{l}^{t} \tag{2.41}$$

$$\mathbf{f}_{2l+1}^{t} = -\mathbf{f}_{2l}^{t} \tag{2.42}$$

$$\mathbf{f}_{2m+1}^{t} = -\mathbf{f}_{2m}^{t} \tag{2.43}$$

where Eq. (2.32) is used to calculate angles α_{lm}^{t} and α_{lm}^{0}.

Here, instead of using a unit vector normal to the anisotropy axis and in the plane where the angle is measured, a unit vector parallel to the second axis is used to reduce computational costs. In order to further reduce computational costs, the angle between two pairs of intersection points is around $90°$ and can be approximated with the cosine of that angle assuming small angle variations during deformation [38].

Using the cosine of the angle instead of the angle in Eqs. (2.40) and (2.41), the torsion spring functions can be rewritten as

$$\mathbf{f}^t_{2\mathrm{l}} = -k_{\mathrm{lm}}(\hat{\zeta}^t_{\mathrm{l}} \cdot \hat{\zeta}^t_{\mathrm{m}} - \hat{\zeta}^0_{\mathrm{l}} \cdot \hat{\zeta}^0_{\mathrm{m}})\hat{\zeta}^t_{\mathrm{m}} \tag{2.44}$$

$$\mathbf{f}^t_{2\mathrm{m}} = -k_{\mathrm{lm}}(\hat{\zeta}^t_{\mathrm{l}} \cdot \hat{\zeta}^t_{\mathrm{m}} - \hat{\zeta}^0_{\mathrm{l}} \cdot \hat{\zeta}^0_{\mathrm{m}})\hat{\zeta}^t_{\mathrm{l}} \tag{2.45}$$

$$\tag{2.46}$$

with noticing that Eqs. (2.42) and (2.43) remain unchanged.

Cubic Torsion Springs

The cubic torsion spring function are:

$$\mathbf{f}^t_{2\mathrm{l}} = -\left(\sum_{i=1}^{3} k_{\mathrm{lm},i}(\alpha^t_{\mathrm{lm}} - \alpha^0_{\mathrm{lm}})^i \right) \hat{\zeta}^t_{\mathrm{m}} \tag{2.47}$$

$$\mathbf{f}^t_{2\mathrm{m}} = -\left(\sum_{i=1}^{3} k_{\mathrm{lm},i}(\alpha^t_{\mathrm{lm}} - \alpha^0_{\mathrm{lm}})^i \right) \hat{\zeta}^t_{\mathrm{l}} \tag{2.48}$$

$$\mathbf{f}^t_{2\mathrm{l}+1} = -\mathbf{f}^t_{2\mathrm{l}} \tag{2.49}$$

$$\mathbf{f}^t_{2\mathrm{m}+1} = -\mathbf{f}^t_{2\mathrm{m}} \tag{2.50}$$

The stiffness parameters for the torsion spring functions can be found by fitting the functions to shear strain-stress curves after converting the spring force to stress using a unit surface.

Using the same approach used for axial springs to avoid scaling the stiffness parameters, the linear torsion spring function can be given by

$$\mathbf{f}^t_{2\mathrm{l}} = -k^t_{\mathrm{lm}}(\alpha^t_{\mathrm{lm}} - \alpha^0_{\mathrm{lm}})\hat{\zeta}^t_{\mathrm{m}} \tag{2.51}$$

$$\mathbf{f}^t_{2\mathrm{m}} = -k^t_{\mathrm{lm}}(\alpha^t_{\mathrm{lm}} - \alpha^0_{\mathrm{lm}})\hat{\zeta}^t_{\mathrm{l}} \tag{2.52}$$

$$\mathbf{f}^t_{2\mathrm{l}+1} = k^t_{\mathrm{lm}}(\alpha^t_{\mathrm{lm}} - \alpha^0_{\mathrm{lm}})\hat{\zeta}^t_{\mathrm{m}} \tag{2.53}$$

$$\mathbf{f}^t_{2\mathrm{m}+1} = k^t_{\mathrm{lm}}(\alpha^t_{\mathrm{lm}} - \alpha^0_{\mathrm{lm}})\hat{\zeta}^t_{\mathrm{l}} \tag{2.54}$$

where

$$k^t_{\mathrm{lm}} = k_{\mathrm{lm}} \, A^t_{\mathrm{j}} \, |\hat{n}^t_{\mathrm{j}} \cdot \hat{\zeta}^t_{\mathrm{l}}| \tag{2.55}$$

2.4.2.2 Deformation Forces using Continuum Mechanics

Ordinary mass-spring systems suffer intrinsically from the problem of spring functions parameterization. The task of parameterization is not straight forward and depends usually on optimization techniques of the parameters values.

Although the parameterization method presented earlier in 2.4.2.1 replaces the optimization process with curves fitting, solving practically the parameterization problem, a big interest in incorporating the calculation of forces via the means of continuum mechanics exists. Because it offers the possibility of calculating the deformation forces using constitutive laws of the modeled materials directly removing the need for parameterization or fitting completely.

For this reason, a method to calculate the deformation forces using continuum mechanics and the theory of finite elasticity was developed. The method, we call the *method of the virtual hexahedron*, can be seen as a way to analytically generate spring functions that incorporate the constitutive laws of the modeled materials. The generated spring functions perform the task of bridging the worlds of continuum mechanics and mass-spring systems .

In the method of the virtual hexahedron, in every volume element, whether a tetrahedron or a hexahedron, a local coordinate system at time $t = 0$ is defined. It is assumed that a deformation tensor describing the transformations of the intersection points also describes the transformations of the volume element's vertices. For that reason a virtual hexahedron with the intersection points in the middle of its surfaces and edges parallel to the three axes is defined (see Fig. 2.13). This virtual hexahedron is used to determine the forces which are applied to the intersection points.

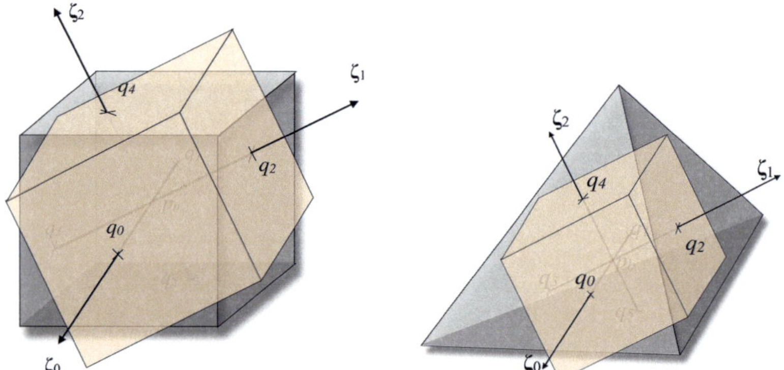

Fig. 2.13 The method of virtual hexahedron, in case of a hexahedral mesh (a), the case of a tetrahedral mesh (b), the intersection points are in the middles of the surfaces of the virtual hexahedron and the edges are parallel to the axes of anisotropies.

Starting with a constitutive law defined by an energy density function W, the forces $\mathbf{f}_i$ which are applied to the intersection points of the virtual hexahedron can be calculated with

$$\mathbf{f}_{21}^t = R_v A_1^0 \mathbf{FBS}^T \mathbf{n}_{21}^0 \tag{2.56}$$

where A_1^0 is the surface of the virtual hexahedron face containing the intersection point q_{21} defined by axis ζ_1 at time $t = 0$, $\mathbf{n}_{21}^0$ is the normal on that face. A_1^0 can be given with

$$A_1^0 = |\zeta_m \times \zeta_n| \quad \text{with} \quad l \neq m \neq n \tag{2.57}$$

R_v is the ratio of the initial volume V^0 of the volume element and the initial volume of the virtual hexahedron V_h^0:

$$R_v = \frac{V^0}{V_h^0} \tag{2.58}$$

$\mathbf{B}$ is the transformation tensor from the local volume element coordinates to the global spatial coordinates. $\mathbf{S}$ is the Piola-Kirchhoff stress tensor and is given with Eq. (1.7), $\mathbf{E}$ is the Green strain tensor which is given with Eq. (1.6)

In Eqs. (2.56) and (1.6), the deformation tensor $\mathbf{F}$ is needed. $\mathbf{F}$ is used to transform a volume element from the undeformed initial state at $t = 0$ to the state at time t. To determine $\mathbf{F}$, two different methods have been implemented:

Aprroximation of the Deformation Tensor $\mathbf{F}$

The first method has been developed to determine an approximation of the deformation tensor using exclusively the strains of axes ζ_1, ζ_2 and ζ_3 in addition to the angles between these axes. The deformation tensor $\mathbf{F}$ can be written as the product of principle strain and shear strain:

$$\mathbf{F} = \begin{pmatrix} \lambda_1 & 0 & 0 \\ 0 & \lambda_2 & 0 \\ 0 & 0 & \lambda_3 \end{pmatrix} \cdot \begin{pmatrix} 1 & \gamma_{12} & \gamma_{13} \\ \gamma_{21} & 1 & \gamma_{23} \\ \gamma_{31} & \gamma_{32} & 1 \end{pmatrix} = \begin{pmatrix} \lambda_1 & \lambda_1\gamma_{12} & \lambda_1\gamma_{13} \\ \lambda_2\gamma_{21} & \lambda_2 & \lambda_2\gamma_{23} \\ \lambda_3\gamma_{31} & \lambda_3\gamma_{32} & \lambda_3 \end{pmatrix} \tag{2.59}$$

The principle strain components λ_i can be approximated using the strain of the axes ζ_i and the angles α_{ij} between each of them (see Fig. 2.14):

$$\beta_{ij} = \beta_{ji} = \frac{1}{2}(\frac{\pi}{2} - \alpha_{ij}) \tag{2.60}$$

$$\lambda_i = \frac{\|\zeta_i\|}{\|\zeta_i^0\|} \prod_{j=1, j\neq i}^{3} \cos(\beta_{ij}) \tag{2.61}$$

and the shear components γ_{ij} are calculated as follows:

$$\gamma_{ij} = \gamma_{ji} = \tan(\beta_{ij}), \quad i \neq j \tag{2.62}$$

Eq. (2.61) is an approximation. Obtaining the exact solution for the principle strain components using the axis strains and the angles is complex. However, if at least one axis is orthogonal to one of the other axis, the exact solution is obtained.

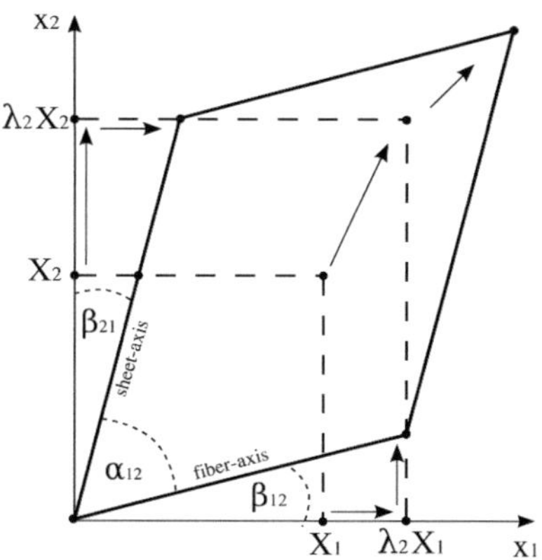

Fig. 2.14 Deformation composed of axial strain and shear strain.

Determining the deformation tensor using shape functions

The second method is a standard procedure known from the finite elements method (FEM). It uses shape functions to calculate a linear interpolation of the deformation tensor from the position of the vertices of the volume element.

The deformation of a linear (8-nodes) hexahedron can be interpolated using shape functions [49, 1]. After transforming the global spatial coordinates of vertices to the local coordinate system of the hexahedron, the transformed coordinates (x_i, y_i, z_i) of the vertices p_i numbered as shown in Figure 2.15 are used to calculate the shape functions N_i according to

$$N_i(\boldsymbol{\xi}) = N_i(\varsigma, \eta, \mu) = \frac{1}{8}(1 + \varsigma_i\varsigma)(1 + \eta_i\eta)(1 + \mu_i\mu) \tag{2.63}$$

And the deformation tensor is given by [49]

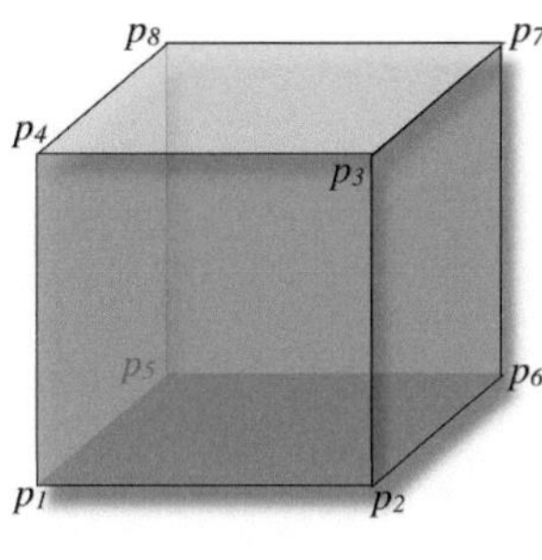

I	ξ_I	η_I	ς_I
1	-1	-1	-1
2	1	-1	-1
3	1	1	-1
4	-1	1	-1
5	-1	-1	1
6	1	-1	1
7	1	1	1
8	-1	1	1

(a)   (b)

Fig. 2.15 Linear hexahedral element and its nodal natural coordinates. Linear hexahedral element (a), nodal natural coordinates for linear hexahedron element (b).

$$\mathbf{F} = \sum_{i=1}^{8} \mathbf{x}_i \otimes \frac{\partial N_i}{\partial \mathbf{X}} \tag{2.64}$$

where

$$\frac{\partial N_i}{\partial \mathbf{X}} = \left(\frac{\partial \mathbf{X}}{\partial \boldsymbol{\xi}} \right)^{-T} \frac{\partial N_i}{\partial \boldsymbol{\xi}} \tag{2.65}$$

$$\frac{\partial \mathbf{X}}{\partial \boldsymbol{\xi}} = \sum_{i=1}^{8} \mathbf{X}_i \otimes \frac{\partial N_i}{\partial \boldsymbol{\xi}} \tag{2.66}$$

and therefore the elements of the deformation tensor are given by

$$F_{IJ} = \sum_{i=1}^{8} x_{i,I} \frac{\partial N_i}{\partial X_J} \tag{2.67}$$

The deformation tensor of a 4-node tetrahedron can be interpolated using shape functions [49, 1]. The local numbering of the vertices of the tetrahedrons is defined by choosing an arbitrary node as the first and then numbering the remaining nodes counterclockwise as seen from the first node in Figure 2.16. As in the hexahedral case, the global spatial coordinates vertices have to be transformed to the local coordinate system. Using the transformed coordinates (x_i, y_i, z_i) of the ordered vertices p_i $(i = 1, \ldots, 4)$ the shape functions N_i are defined by

$$\mathbf{X} = \begin{pmatrix} 1 \\ X \\ Y \\ Z \end{pmatrix} = \mathbf{A} \begin{pmatrix} N_1 \\ N_2 \\ N_3 \\ N_4 \end{pmatrix} \tag{2.68}$$

where

$$\mathbf{A} = \begin{pmatrix} 1 & 1 & 1 & 1 \\ x_1 & x_2 & x_3 & x_4 \\ y_1 & y_2 & y_3 & y_4 \\ z_1 & z_2 & z_3 & z_4 \end{pmatrix} \tag{2.69}$$

By inverting matrix $\mathbf{A}$, the shape functions N_i can be written as a linear combination of the components of $\mathbf{X}$ with

$$N_i(x, y, z) = m_{1,i} + m_{2,i}X + m_{3,i}Y + m_{4,i}Z \tag{2.70}$$

The elements m_{IJ} of $\mathbf{A}^{-1}$ can be calculated using the following relation:

$$m_{IJ} = (-1)^{I+J} \frac{1}{\det(\mathbf{A})} \hat{\mathbf{A}}_{IJ} \tag{2.71}$$

where $\hat{\mathbf{A}}_{IJ}$ are the minors of $\mathbf{A}$. Finally, the tetrahedron deformation tensor can then be given by

$$F_{IJ} = \sum_{i=1}^{4} x_{i,I} \frac{\partial N_i}{\partial X_J} \tag{2.72}$$

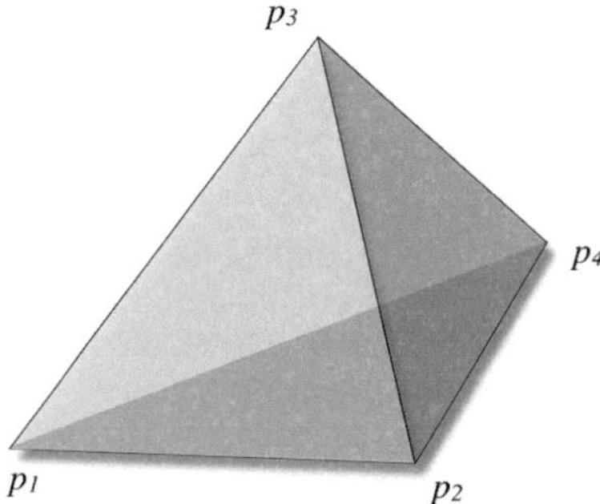

Fig. 2.16 A 4-node tetrahedral element and the nodes numbering scheme

Unlike the approximation method described above, this method is well-established and has less computational costs. Therefore, it is used as the method of choice for calculating the deformation tensor. Back to Eq. (2.56):

$$\mathbf{f}_{21}^{t} = R_{v}A_{1}^{0}\mathbf{FBS}^{T}\mathbf{n}_{21}^{0}$$

$\mathbf{S}$ is defined for the local coordinate system in the volume element, the normal vectors $\mathbf{n}_{21}^{0}$ on the surfaces of the virtual hexahedron are:

$$
\begin{aligned}
\mathbf{n}_{0}^{0} = -\mathbf{n}_{1}^{0} = \begin{pmatrix} 1 \\ 0 \\ 0 \end{pmatrix} = \mathbf{B}^{-1}\hat{\zeta}_{1}^{0} \\
\mathbf{n}_{2}^{0} = -\mathbf{n}_{3}^{0} = \begin{pmatrix} 0 \\ 1 \\ 0 \end{pmatrix} = \mathbf{B}^{-1}\hat{\zeta}_{2}^{0} \\
\mathbf{n}_{4}^{0} = -\mathbf{n}_{5}^{0} = \begin{pmatrix} 0 \\ 0 \\ 1 \end{pmatrix} = \mathbf{B}^{-1}\hat{\zeta}_{3}^{0}
\end{aligned}
\tag{2.73}
$$

Using these normal vectors and that $\mathbf{S} = \mathbf{S}^{T}$, the force $\mathbf{f}_{21}$ acting on intersection point q_j can be decomposed in three force components along the axes. Substituting in Eq. (2.56) and using $\lambda = l + 1$ gives:

$$
\begin{aligned}
\mathbf{f}_{21} &= R_{v}A_{1}^{0}\mathbf{FBS}\mathbf{n}_{21}^{0} \\
&= R_{v}A_{1}^{0}\mathbf{FB}(S_{1\lambda}\mathbf{n}_{1}^{0} + S_{2\lambda}\mathbf{n}_{3}^{0} + S_{3\lambda}\mathbf{n}_{5}^{0}) \\
&= R_{v}A_{1}^{0}\mathbf{FB}\left(S_{1\lambda}\mathbf{B}^{-1}\hat{\zeta}_{1}^{0} + S_{2\lambda}\mathbf{B}^{-1}\hat{\zeta}_{2}^{0} + S_{3\lambda}\mathbf{B}^{-1}\hat{\zeta}_{3}^{0}\right) \\
\mathbf{f}_{21} &= R_{v}A_{1}^{0}\left(S_{1\lambda}\hat{\zeta}_{1} + S_{2\lambda}\hat{\zeta}_{2} + S_{3\lambda}\hat{\zeta}_{3}\right)
\end{aligned}
\tag{2.74}
$$

which is a function suitable for code implementation in contrast to Eq. (2.56).

In comparison to the concept of axial and angular springs, the terms $A_{1}^{0}S_{11}\hat{\zeta}_{1}$, $A_{2}^{0}S_{22}\hat{\zeta}_{2}$ and $A_{3}^{0}S_{33}\hat{\zeta}_{3}$ represent the forces of axial springs, while the remaining terms represent components of torsion spring forces.

2.4.2.3 Volume Preservation

Developing a method for volume preservation was a very important part of this work, due to high interest in modeling objects that retain constant volume during deformation, like cardiac myocytes.

Bourguignon *et al.* [38] used volume preservation springs acting on the volume elements particles. The function Bourguignon used to calculate the volume preservation force at vertex p_i of a hexahedron is given with

$$\mathbf{f}_i^t = -\left(k_s(\|\boldsymbol{\xi}_i^t\| - \|\boldsymbol{\xi}_i^0\|) + k_d\boldsymbol{\vartheta}_i^t \cdot \hat{\boldsymbol{\xi}}_i^t\right)\hat{\boldsymbol{\xi}}_i^t \tag{2.75}$$

In case of using tetrahedrons, the function is given with

$$\mathbf{f}_i^t = -k_s\left(\sum_{j=1}^{4}\|\boldsymbol{\xi}_j^t\| - \sum_{j=1}^{4}\|\boldsymbol{\xi}_j^0\|\right)\hat{\boldsymbol{\xi}}_i^t \tag{2.76}$$

where $\boldsymbol{\xi}_i^t = \mathbf{x}_b^t - \mathbf{x}_i^t$, and $\mathbf{x}_b$ is the barycenter of the volume element and can be calculated using Eq. (2.8) for hexahedrons and Eq. (2.22) for tetrahedrons. The unit vector $\hat{\boldsymbol{\xi}}_i^t$ is given by

$$\hat{\boldsymbol{\xi}}_i^t = \frac{\boldsymbol{\xi}_i^t}{\|\boldsymbol{\xi}_i^t\|} \tag{2.77}$$

and for the relative velocity of an intersection point ϑ^t we can write:

$$\boldsymbol{\vartheta}_i^t = \frac{d\mathbf{x}_i^t}{dt} - \frac{d\mathbf{x}_b^t}{dt} \tag{2.78}$$

These volume preservation springs exert penalty forces on particles proportional to the strain of the springs.

This approach did not lead to satisfying results. The variations of volume of the elements were too high. The parameterization of the functions was an additional drawback of this method.

An adaptive volume preservation spring function was proposed with the hope of obtaining a better volume preservation than achieved using the Bourguignon spring functions in Eqs. (2.75) and (2.76).

The main idea behind the adaptive spring functions method is to make the value k_s in Eq. (2.76) adaptive with time, and to include an adaptation algorithm that continuously updates the value of k_s to reduce the volume difference ΔV between the original and the current volume of the element, V^0 and V^t respectively:

$$\Delta V = V^t - V^0 \tag{2.79}$$

We will use the symbol $\hat{k}_s$ to distinguish the adaptive k_v from the constant counterpart.

The Least Mean Squares (LMS) algorithm was used to update the value of k_s, using ΔV as an adaptation variable. For example for a tetrahedron:

$$\mathbf{f}_i^t = -\hat{k}_s^{\;t}\left(\sum_{j=1}^{4}\|\boldsymbol{\xi}_j^t\| - \sum_{j=1}^{4}\|\boldsymbol{\xi}_j^0\|\right)\hat{\boldsymbol{\xi}}_i^t \tag{2.80}$$

$$\hat{k}_s^{\;t+h} = \hat{k}_s^{\;t} + \mu\Delta V\sum_{j=1}^{4}\|\boldsymbol{\xi}_j^t\| \tag{2.81}$$

where $\hat{k}_s^{\;t}$ is the current adaptive stiffness, $\hat{k}_s^{\;t+h}$ is the adaptive stiffness after a time step h, μ is the adaptation constant or the LMS step size.

Although this adaptive springs stiffness showed promising results in initial experiments with a small number of elements, this method was not further investigated.

To enforce volume preservation in each element of the model, a method used in continuum mechanics is finally applied to our mass-spring system. According to continuum mechanics, volume preservation can be introduced to an element by adding a volumetric energy density term W_v to the potential energy density function of the element W [49, 1]:

$$\hat{W} = W + W_v \tag{2.82}$$

W_v depends on the change of volume and can be formulated using the ratio of the current to the initial volume. There are many possible ways to formulate W_v. In this work the following form was used:

$$W_v = p(\frac{\Delta V}{V^0})^2 \tag{2.83}$$

We know that the determinant of the deformation tensor $\mathbf{F}$ of a an element is

$$\det(\mathbf{F}) = \frac{V^t}{V^0} \tag{2.84}$$

and we can then write

$$\frac{\Delta V}{V^0} = \frac{V^t - V^0}{V^0} = \det(\mathbf{F}) - 1 \tag{2.85}$$

By substituting Eq. (2.85) in Eq. (2.83) we get

$$W_v = p(\det(\mathbf{F}) - 1)^2 \tag{2.86}$$

The volumetric term W_v represents the potential energy resulting from a change in volume.

A common method to achieve incompressibility is to minimize the total energy $\hat{W}$ with respect to the displacement. In that case a variation calculus problem with constraints has to be solved, and the coefficient p can be seen as a Lagrange multiplier [49, 1].

In this work, a different approach adapted to the mass-spring system is used. Here, volume preservation penalty forces, that depend on the ΔV, are derived from a Piola-Kirchhoff tensor $\mathbf{S}$ of volume preservation which is in turn derived from the volumetric energy density term W_v according to

$$S_{ij} = \frac{\partial W_v}{\partial E_{ij}} \tag{2.87}$$

where E_{ij} are the Green strain tensor elements.

To compute S_{ij}, W_v must be transformed from a function of the deformation tensor elements F_{ij} to a function of the Green strain tensor elements E_{ij}. By expanding W_v:

$$W_v = p(\det(\mathbf{F}) - 1)^2 = p(\det(\mathbf{F})^2 - 2\det(\mathbf{F}) + 1) \tag{2.88}$$

In general, given two matrices $\mathbf{A}$ and $\mathbf{B}$, one can write:

$$\det(\mathbf{A}) = \det(\mathbf{A}^T) \tag{2.89}$$
$$\det(\mathbf{AB}) = \det(\mathbf{A})\det(\mathbf{B}) \tag{2.90}$$
$$\det(\mathbf{AA}) = \det(\mathbf{A})\det(\mathbf{A}) \tag{2.91}$$
$$= \det(\mathbf{A})\det(\mathbf{A}^T) \tag{2.92}$$
$$= \det(\mathbf{AA}^T) = \det(\mathbf{A})^2 \tag{2.93}$$

The Green strain tensor and the deformation tensor are related as follows:

$$\mathbf{E} = \frac{1}{2}(\mathbf{FF}^T - \mathbf{I}) \Rightarrow \tag{2.94}$$
$$\mathbf{FF}^T = 2\mathbf{E} + \mathbf{I} \tag{2.95}$$

Using Eq. (2.93) and Eq. (2.95) we obtain:

$$\det(\mathbf{F})^2 = \det(\mathbf{FF}^T) = \det(2\mathbf{E} + \mathbf{I}) \tag{2.96}$$

and Eq. 2.88 can be rewritten as

$$W_v = p(\det(2\mathbf{E} + \mathbf{I}) - 2\sqrt{\det(2\mathbf{E} + \mathbf{I})} + 1) \tag{2.97}$$

$$= p(\Omega - 2\sqrt{\Omega} + 1) \tag{2.98}$$

where

$$
\begin{aligned}
\Omega = \det(2\mathbf{E} + \mathbf{I}) = &(2E_{11} + 1)(2E_{22} + 1)(2E_{33} + 1) \\
&+ 8(E_{12}E_{23}E_{31}) + 8(E_{21}E_{32}E_{13}) \\
&- 4(E_{23}E_{32}(2E_{11} + 1)) \\
&- 4(E_{13}E_{31}(2E_{22} + 1)) \\
&- 4(E_{12}E_{21}(2E_{33} + 1))
\end{aligned}
\tag{2.99}
$$

and finally, the elements of the tensor $\mathbf{S}$ can be given with

$$S_{ij} = p\frac{\partial \Omega}{\partial E_{ij}}\left(1 - \frac{1}{\sqrt{\Omega}}\right) \tag{2.100}$$

where

$$
\frac{\partial \Omega}{\partial E_{ij}} =
\begin{cases}
2(2E_{jj} + 1)(2E_{kk} + 1) - 8E_{kj}E_{jk} & i = j \\
8(E_{jk}E_{ki}) - 4E_{ji}(2E_{kk} - 1) & i \neq j
\end{cases}
$$

The resulting volume preservation forces can be calculated using the method of virtual hexahedron represented by Eq. (2.56) or Eq. (2.74) and the volume preservation Piola-Kirchhoff tensor elements S_{ij} can be calculated with Eq. (2.100).

If the volume preservation effect should act in the direction of the defined anisotropies in an element, these forces can be applied to the intersections points and then distributed to the vertices using Eq. (2.29). In the other case where the volume preservation effect should be isotropic, a coefficient matrix C that guarantees an equal share of volume preservation forces to all vertices of an element is generated and used in Eq. (2.29) to distribute the forces to the vertices.

It is important to mention that some constitutive laws include terms to enforce volume preservation. In that case using the constitutive law with the method of the virtual hexahedron to calculate the deformation forces is enough to ensure the volume preservation, meaning that no additional separate treatment is needed.

2.4.2.4 Volume Control

In some applications a method to control the volume of the element is desired (e.g. see Section 5.4.3). In these applications the volume of the element should not remain constant, but rather change towards a predefined value. To add that possibility to our framework, the method used to enforce volume preservation by adding a volume preservation energy density term was improved to allow for volume control.

Starting with Eq. (2.86), one can write:

$$W_{\mathrm{v}} = p(\det(\mathbf{F}) - r)^2 \tag{2.101}$$

where r is the ratio of the target volume V^∞ to the initial volume V^0 of the element:

$$r = \frac{V^\infty}{V^0} \tag{2.102}$$

By setting the target volume to the initial volume in Eq. (2.102) we get $r = 1$, turning Eq. (2.101) back to Eq. (2.86).

By following the same procedure used to derive the volume preservation forces, we can obtain the equations for the volume control forces. We start by expanding Eq. (2.101):

$$W_{\mathrm{v}} = p(\det(\mathbf{F})^2 - 2r\ \det(\mathbf{F}) + r^2) \tag{2.103}$$

Then by substituting the terms of the deformation tensor $\mathbf{F}$ with the Green strain tensor $\mathbf{E}$ using the Eq. (2.96) we get:

$$
\begin{aligned}
W_{\mathrm{v}} &= p(\det(2\mathbf{E} + \mathbf{I}) - 2r\sqrt{\det(2\mathbf{E} + \mathbf{I})} + r^2) \tag{2.104}\\
&= p(\Omega - 2r\sqrt{\Omega} + r^2) \tag{2.105}
\end{aligned}
$$

where Ω is given in Eq. (2.99). Finally the Piola-Kirchhoff stress tensor elements can be calculated using

$$S_{\mathrm{ij}} = p\frac{\partial \Omega}{\partial E_{\mathrm{ij}}}\left(1 - \frac{r}{\sqrt{\Omega}}\right) \tag{2.106}$$

and the resulting intersection points forces can be calculated using the method of virtual hexahedron represented by Eq. (2.56) or Eq. (2.74).

2.4.2.5 Material Specific Friction

Anisotropic friction specific for the different material in each volume element of the modeled object was implemented using the relative velocity of the intersection points:

$$\dot{\zeta}_{\mathrm{l}}^{t} = \frac{d}{dt}(\mathbf{x}_{2\mathrm{l}}^{t} - \mathbf{x}_{2\mathrm{l}+1}^{t}) \tag{2.107}$$

Bourguignon *et al.* introduced linear friction by applying friction forces to the intersection points using the following function:

$$\mathbf{f}_{21}^t = - k_{\mathrm{d}}(\dot{\zeta}_1^t \cdot \hat{\zeta}_1^t)\hat{\zeta}_1^t \tag{2.108}$$

where k_{d} is the friction coefficient, or as Bourguignon calls it, the damping parameter of spring S_1. This kind of friction was also used in the work of M. Mohr [39] for anisotropy springs in his hybrid-mass-spring model.

In our model, the following function was used to introduce linear friction:

$$\mathbf{f}_{j}^t = - k_{\mathrm{d}}^t(\dot{\zeta}_1^t \cdot \hat{\zeta}_1^t)\hat{\zeta}_1^t \tag{2.109}$$

where k_{d}^t is given by

$$k_{\mathrm{d}}^t = k_{\mathrm{d}}\, S_j^t\, |\hat{n}_j^t \cdot \hat{\zeta}_1^t| \tag{2.110}$$

S_j^t is the surface of the face containing intersection point q_j that the axis ζ_1 defines and n_j^t is the normal on that face. When using the method of the virtual hexahedron the following formula can be used:

$$\mathbf{f}_{21}^t = - R_{\mathrm{v}}(k_{\mathrm{d}}\dot{\zeta}_1^t \cdot \hat{\zeta}_1^t)A_1^0\hat{\zeta}_1^t \tag{2.111}$$

where A_1^0 is the surface of the virtual hexahedron face containing q_{21} of axis ζ_1 defined at time $t = 0$. A_1^0 can be calculated using Eq. (2.57).

2.4.3 Active Forces

To model tensions generated by internal processes, like the contraction tension in a muscle along its fibers, mathematical models of these internal processes can be used to calculate the tension these processes generate in each volume element. These tensions can also be measured using special measurement techniques and then introduced to the mechanical modeling framework. Depending on the process, the resulting force might be isotropic affecting all particles of the volume element in the same way, or anisotropic. In the isotropic case, the tension $\mathbf{t}_j$ calculated for each face of the volume element is transformed to forces using the surface of the face S_j and distribute it equally to the particles at the vertices p_i of the face S_j:

$$\mathbf{f}_i = \frac{1}{n}\mathbf{t}_j\mathbf{n}_j \tag{2.112}$$

where n is the number of vertices defining the face, and $\mathbf{n}_j$ is the normal on that face. In the anisotropic case, tension $\mathbf{t}_l$ generated along the axis ζ_l is set to the intersection points defined by the axis with

$$\mathbf{f}_{2l} = \frac{1}{2}\mathbf{t}_l S_{2l}^t \hat{\zeta}_l \tag{2.113}$$

$$\mathbf{f}_{2l+1} = -\frac{1}{2}\mathbf{t}_l S_{2l+1}^t \hat{\zeta}_l \tag{2.114}$$

where, S_{2l}^t is the surface of the face where axis ζ_l defines the intersection point q_{2l}, and S_{2l}^t is the surface of the face where the same axis defines the second intersection point q_{2l+1}, both at time t. And when using the method of virtual hexahedron we get:

$$\mathbf{f}_{2l} = \frac{1}{2}R_v \mathbf{t}_l A_{2l}^t \hat{\zeta}_l^t \tag{2.115}$$

$$\mathbf{f}_{2l+1} = -\frac{1}{2}R_v \mathbf{t}_l A_{2l+1}^t \hat{\zeta}_l^t \tag{2.116}$$

where A_{2l}^t and A_{2l+1}^t are the surfaces of the virtual hexahedron faces containing intersection points q_{2l} and q_{2l+1} respectively, that axis ζ_l defines computed at time t.

Internal processes that generate force and not tension, like the myocardial tension development where the count of myofibrils is independent of the deformation, the initial surface S^0 (in spring function formulation) or A^0 (in the virtual hexahedron method formulation) is used instead of the surface at time t to calculate the force resulting from the internal process.

2.4.4 External Forces

In contrast with internal forces, external forces are set directly to mass particles. They do not depend on the state of volume elements to which these particles belong. These forces can be part of the environment like gravity forces, or part of the simulation setup, like initial velocity.

2.4.4.1 External Acceleration

If the application should require the setting of external acceleration to the mass particles, this can be done by calculating the forces that correspond to the acceleration and assign the forces to the particles. For a particle p_i, the force resulting from an external acceleration $\mathbf{a}$ can be given according to Newton's second law of motion:

$$\mathbf{f}_i = m_i \mathbf{a} \tag{2.117}$$

where m_i is the mass of p_i .

Loading the particles with gravity is just a special case of the general external acceleration where the acceleration vector $\mathbf{a}$ is the gravitational acceleration vector $\mathbf{g}$.

2.4.4.2 Global Friction

Linear friction can also be introduced to the system using a friction coefficient $\mu \geq 0$. When $\mu = 0$ no friction is modeled. In the case $\mu > 0$, each particle p_i becomes subject to a global friction force $\mathbf{f}_i$ given with

$$\mathbf{f}_i = -\mu \mathbf{v}_i \tag{2.118}$$

where $\mathbf{v}_i$ is the velocity of p_i.

2.4.4.3 External Pressure

Constant or time dependent pressure P^t can be set in the defined cavities. The pressure is transformed to forces that affect vertices of the cavity faces. For a face $\mathcal{F}_j$ defined by vertices p_i $(i = 1, \ldots, n)$ the following rule is used:

$$\mathbf{f}_i = -\frac{1}{n} P^t A_j^t \mathbf{n}_j \tag{2.119}$$

where A_j^t is the surface of the face $\mathcal{F}_j$ at time t and $\mathbf{n}_j$ is the normal on the face pointing towards the cavity.

2.4.4.4 Initial Velocity

Many mechanical modeling applications require giving the modeled object an initial velocity right at the beginning of the simulation. This can be done by setting the velocity vectors of all particles directly to the chosen initial velocity vector $\mathbf{v}^0$ which is a $3n$ vector and n is the number of particles.

2.5 Time Integration

In this section, the equations of motion and the resulting system of coupled ordinary differential equations (ODE)s, implemented schemes for numerical time integration and several other topics related to time integration are presented.

2.5.1 Equations of Motion

The coordinates $\mathbf{x}_i = (x_i, y_i, z_i)$ of particles p_i ($i = 1, \ldots, n$), where n is the count of all particles of the object, along with the coordinates $\mathbf{x}_{i,\text{init}}$ describe the deformation state of the object at any given time t. Using $\mathbf{x}_i$, we can build the $3n$ dimensional vector $\mathbf{u}$:

$$\mathbf{u} = (x_1, y_1, z_1, x_2, y_2, z_2, \ldots, x_n, y_n, z_n) \tag{2.120}$$

$\mathbf{u}$ describes the trajectory of all particles. Similarly, we can define the initial trajectories vector $\mathbf{u}_{\text{init}}$.

The particles' motion is defined by a set of $3n$ coupled second-order differential equations:

$$\mathbf{M}\frac{d^2}{dt^2}\mathbf{u} + \mathbf{F}(\mathbf{u}, \mathbf{u}_{\text{init}}, \frac{d}{dt}\mathbf{u}, t) = 0 \tag{2.121}$$

with the boundary conditions:

$$\mathbf{u}\,|_{(t=0)} = \mathbf{u}^0$$

$$\frac{d}{dt}\mathbf{u}\,|_{(t=0)} = \mathbf{v}\,|_{(t=0)} = \mathbf{v}^0$$

where

$$\mathbf{F}(\mathbf{u}, \mathbf{u}_{\text{init}}, \frac{d}{dt}\mathbf{u}, t) = \mathbf{F}_\mu(\frac{d}{dt}\mathbf{u}) + \mathbf{F}_d(\mathbf{u}, \mathbf{u}_{\text{init}}) + \mathbf{F}_a(t) \tag{2.122}$$

$$\tag{2.123}$$

and $\mathbf{M}$ is the system's $3n \times 3n$ mass tensor, $\mathbf{F}_\mu$ expresses friction forces, $\mathbf{F}_d$ expresses passive forces resulting from the deformation of the body and $\mathbf{F}_a$ the time-dependent active forces. Here, a distigtion between the initial trajectories $\mathbf{u}_{\text{init}}$ resulting during the structure initialization phase, and $\mathbf{u}^0$ which is the trajectories vector at $t = 0$ is made. $\mathbf{u}^0$ could differ from $\mathbf{u}_{\text{init}}$ if an initial displacement $\Delta\mathbf{u}$ was introduced to the system:

$$\mathbf{u}_{\text{init}} = \mathbf{u}^0 + \Delta\mathbf{u} \tag{2.124}$$

$\mathbf{u}_{\text{init}}$ does not play a role in the derivations of the following equations, therefore it will be ommited from the notation. The nature of these boundary conditions makes the ordinary differential equation Eq. (2.121) an *initial value problem*.

Every ordinary differential equation of order m can be separated into a system of m coupled partial differential equation of first order. According to this, the motion equations of the particles can be rewritten as $2 \times 3n$ first-order equations:

$$\frac{d}{dt}\mathbf{u} = \mathbf{v} \tag{2.125}$$

$$\frac{d}{dt}\mathbf{v} = \mathbf{M}^{-1}\mathbf{F}(\mathbf{u}, \frac{d}{dt}\mathbf{u}, t) \tag{2.126}$$

Time integration is solving these equations for $\mathbf{u}$.

There are several methods for numerical time integration of initial value problem ODEs. However, the underlying idea of any of these numerical methods is rewriting $d\mathbf{u}$ and dt as finite steps $\Delta\mathbf{u}$ and Δt, and multiply the equation by Δt derivatives in the form of differences. The smaller Δt is, the better approximation of the differential equation can be achieved. The *forward Euler* method is a literal implementation of this procedure. The implementation of the forward Euler method is discussed in the following.

2.5.2 Explicit Euler Method

The Euler method is a first order method for solving ODEs. The function $u(t)$ is evaluated at equidistant time points t_i where $t_{i+1} = t_i + h$ and h is the time step:

$$\mathbf{v}\,|_{(t=t_i)} = \mathbf{v}^0 + \sum_{k=0}^{i-1} \int_{t_k}^{t_k+h} \mathbf{M}^{-1}\mathbf{F}(\mathbf{u}, \mathbf{v}, t)dt \tag{2.127}$$

The integration can be approximated by

$$\int_{t_k}^{t_k+h} \mathbf{M}^{-1}\mathbf{F}(\mathbf{u}, \mathbf{v}, t)dt = \mathbf{M}^{-1}\mathbf{F}(\mathbf{u}_k, \mathbf{v}_k, t)h \tag{2.128}$$

using this approximation, the Euler steps can be obtained as follows:I

$$\mathbf{v}^{i+1} = \mathbf{v}^i + \mathbf{M}^{-1}\mathbf{F}(\mathbf{u}^i, \mathbf{v}^i, t_i)h \tag{2.129}$$

$$\mathbf{u}^{i+1} = \mathbf{u}^i + \mathbf{v}^i h \tag{2.130}$$

The magnitude of the errors that occur between Euler step $\mathbf{v}^n$ and $\mathbf{v}^{n+1}$ can be estimated by making a comparison with a Taylor expansion. By assuming $\mathbf{v}(t_n)$ is exactly known, the Taylor expansion gives:

$$\mathbf{v}\,|_{(t=t_i+h)} = \mathbf{v}\,|_{(t=t_i)} + h\frac{d}{dt}\mathbf{v}\,|_{(t=t_i)} + O(h^2) + O(h^3) + \dots \tag{2.131}$$

$$= \mathbf{v}^{i+1} + O(h^2) + O(h^3) + \dots \tag{2.132}$$

For small h the error resulting from the Euler method is proportional to h^2.

2.5.3 Stiff Differential Equations

When using the forward Euler method as well as many other methods, h must be chosen sufficiently small to obey an absolute stability restriction. The absolute stability restriction ensures that the distance between the exact and numerical solutions decreases. Otherwise, the algorithm becomes unstable. It is important to understand that the absolute stability restriction does not ensure accuracy [55].

Ideally, the choice of the time step h should be dictated by the approximation accuracy requirement. But the additional absolute stability requirement can dictate a much smaller h [55].

Loosely speaking, differential equations of an initial value problem are called stiff, if the absolute stability requirements of the equations dictates a much smalled h than the approximation requirements demands [55].

Generally, mass-spring systems generate stiff differential equations. That applies for Adamss as well. In almost all the applications where Adamss was implemented the resulting differential equations turned out to be stiff for a fairly small time step. That means that even smaller time steps were needed to ensure the stability of the time integration algorithm. That also means much more integration steps were needed to obtain the results of a simulation, making the system very expensive computationally.

*****

Actually, the integration algorithms became unstable so many times while trying to find the biggest possible time step the numerical time integration can take without becoming unstable. And so many times, the unstable system produced impossible deformations that could be elevated to the status of computational art. In an attempt to find beauty in failures. Figure 2.17 depicts a case of unphysical deformation. Because of the stiffness problem, the saying: "My mood is oscillating like a Mass-Spring System" made perfect sense.

*****

Stiffness of ODEs is a well known issue in numerical integration, and much effort has been done in the development of methods that try to extend the region of stability of the differential equations. To deal with the stiffness problem, two approaches were implemented in this work. The first is the *Adams-Bashforth Moulton predictor-corrector* method with adaptive time- stepping, and the second is the *backward Euler* method which is an *implicit* method. The computational cost of methods that solve the stiffness problem is usually high. But, overall the use of

Fig. 2.17 A Snapshots of a simulations right before (left) and right after (right) time integration of the stiff ODEs became unstable.

these methods may be much cheaper than taking the large number of steps required for explicit methods to work successfully [56].

2.5.4 Adams-Bashforth Moulton Predictor-Corrector Method

To be able to take larger time-steps, two multistep methods, namely *Adams-Bashforth* which is an explicit method and *Adams-Moulton* which is an implicit methods were used as a *Predictor-Corrector* pair with adaptive time-stepping as described in [57].

The explicit Adams-Bashforth method of order $k + 1$ is given for $\mathbf{u}$ with

$$\mathbf{u}^{i+1} = \mathbf{u}^i + h \sum_{j=0}^{k} \beta_j \mathbf{v}^{i-j} \tag{2.133}$$

$$\beta_j = (-1)^j \sum_{i=j}^{k} \binom{i}{j} \gamma_i \tag{2.134}$$

$$\gamma_i = (-1)^i \int_0^1 \binom{-s}{i} ds \tag{2.135}$$

$$\tag{2.136}$$

and the different values for β_j can be also found in tables in the literature [58, 57]. Adams-Bashforth methods are explicit methods with very small regions of absolute stability. The implicit version of Adams methods are called the Adams-Moulton methods of order k+2:

$$\mathbf{u}^{i+1} = \mathbf{u}^i + h \sum_{j=-1}^{k} \beta_j \mathbf{v}^{i-j} \tag{2.137}$$

To make use of both methods in a predictor-corrector pair, the explicit Adams-Bashforth method is used to predict the value of the next step. By using the predicted value on the right side of the implicit Adams-Moulton method, a better estimation of the next step is made, i.e. the value is corrected, as presented in these formulas where first a prediction is made using the Adams-Bashforth method of the 3rd order:

$$\tilde{\mathbf{v}}^{i+1} = \mathbf{v}^i + h \left(\frac{23}{12} \mathbf{M}^{-1} \mathbf{F}(\mathbf{u}^i, \mathbf{v}^i, t_i) - \frac{16}{12} \mathbf{M}^{-1} \mathbf{F}(\mathbf{u}^{i-1}, \mathbf{v}^{i-1}, t_{i-1}) \right.$$
$$\left. + \frac{5}{12} \mathbf{M}^{-1} \mathbf{F}(\mathbf{u}^{i-2}, \mathbf{v}^{i-2}, t_{i-2}) \right) \tag{2.138}$$

$$\tilde{\mathbf{u}}^{i+1} = \mathbf{u}^i + h \left(\frac{23}{12} \mathbf{v} - \frac{16}{12} \mathbf{v}^{i-1} + \frac{5}{12} \mathbf{v}^{i-2} \right) \tag{2.139}$$

Then, the resulting values are used in the Adams-Moulton method of the same order in the correction step:

$$\mathbf{v}^{i+1} = \mathbf{v}^i + h \left(\frac{5}{12} \tilde{\mathbf{v}}^{i+1} + \frac{8}{12} \mathbf{M}^{-1} \mathbf{F}(\mathbf{u}^i, \mathbf{v}^i, t_i) - \frac{1}{12} \mathbf{M}^{-1} \mathbf{F}(\mathbf{u}^{i-1}, \mathbf{v}^{i-1}, t_{i-1}) \right) \tag{2.140}$$

$$\mathbf{u}^{i+1} = \mathbf{u}^i + h \left(\frac{5}{12} \tilde{\mathbf{u}}^{i+1} + \frac{8}{12} \mathbf{v}^i - \frac{1}{12} \mathbf{v}^{i-1} \right) \tag{2.141}$$

The predicted and the corrected coordinates are compared, and a decision about the accuracy at the specified time-step is taken. If all the differences were smaller than a predefined lower-error threshold, the time-step is doubled (*reducing time-mesh*). If one of the differences was greater than a predefined maximal-error threshold, the time-step is divided by two (*refining time-mesh*). Otherwise, the time-step remains unchanged. The method is depicted in the flowchart 2.18.

Adams-Bashforth and Adams-Moulton methods are multi-points methods, old values should be available to the time integration method. These values must be taken by equidistant time-steps. When the time-step is halved, extra values must be calculated between the old ones. To do so the old values are linearly interpolated. The

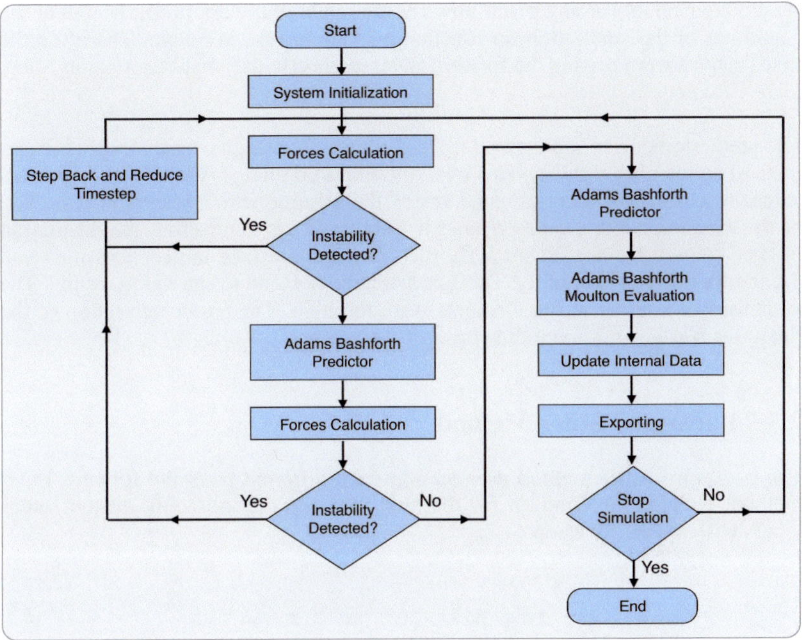

Fig. 2.18 Flowchart of the Adams-Bashforth-Moulton time integration scheme, the time step adaptation mechanism occurs in the Adams Bashforth Moulton evalutation block

refine time-mesh concept is described in Figure 2.19.

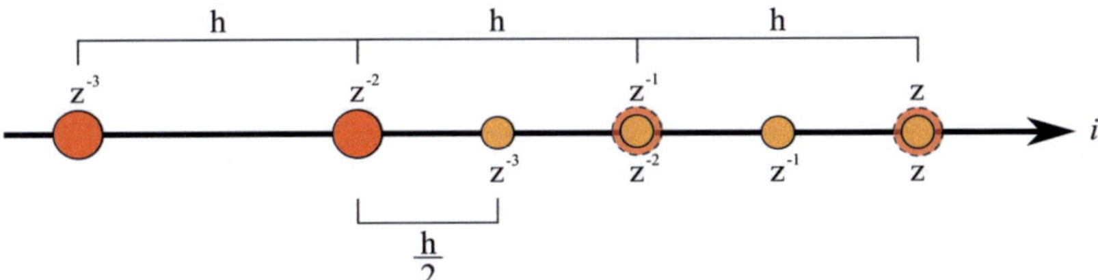

Fig. 2.19 The time mesh refinement procedure is displayed, i is the discrete time axis, the big circles represent the old values, they are at equidistance of h the old timestep, the little circles represent the new values aligned with the new time-step $\frac{h}{2}$, new values at Z^{-3} and Z^{-1} are linearly interpolated with the old values at Z, Z^{-1} and Z^{-2}.

The implemented method is of the 3rd order. That means that three previous points should be available at any given time for the method to work properly. But at the beginning of the simulation no function evaluations are available. Therefore the first 2 steps are done using the forward Euler method with a small enough timestep.

The prediction-correction method offered a huge advantage over the explicit Euler method concerning stability. However, the timestep plunges often into very small values to ensure stability and accuracy of the solution which results in reduction of the efficiency. On the other hand if the thresholds controlling the adaptation of time steps were not set properly the system could take larger steps that lead eventually towards instability. This approach did not lead to satisfying results. The variations of volume of the elements were too high. The parameterization of the functions was an additional drawback of this method.

2.5.5 Backward Euler Method

The backward Euler method uses an approach different from the forward Euler method. Hereby, the solution for the next timestep depends only on the forces which will arise at timestep t_{n+1}:

$$\mathbf{v}^{i+1} = \mathbf{v}^i + \Delta\mathbf{v} \tag{2.142}$$

$$\Delta\mathbf{v} = \mathbf{M}^{-1}F(\mathbf{u}^{i+1}, \mathbf{v}^{i+1}, t_{i+1})h \tag{2.143}$$

$$\mathbf{u}^{i+1} = \mathbf{u}^i + \mathbf{v}^{i+1} \cdot h \tag{2.144}$$

The backward Euler method is an implicit method, since a set of implicit equations has to be solved.

Figure 2.20 shows the exact solution for a differential equation of a harmonic oscillation and the approximated solutions of the explicit and implicit Euler methods for different time step sizes. Depending on the time step size, the amplitude of the solution resulting from the explicit Euler method grows in time without bound, which leads to numerical instability. By contrast, the solution of the backward Euler method shows numerical damping, which is characteristical for implicit methods allowing the use of bigger time steps.

By applying a Taylor series expansion to $\mathbf{F}(\mathbf{u}^{i+1}, \mathbf{v}^{i+1}, t_{i+1})$ and making the first order approximation, we get:

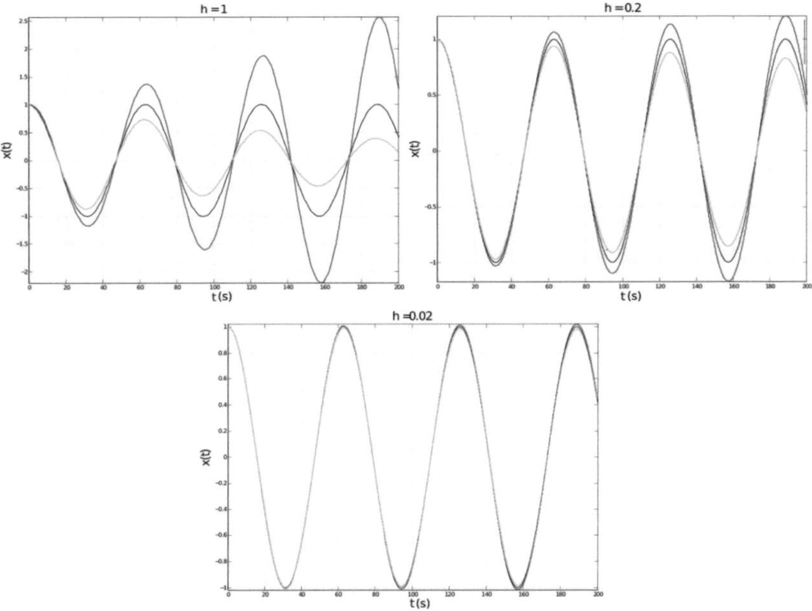

Fig. 2.20 Comparison of explicit and implicit method for different time step sizes. The exact solution is displayed in blue, the explicit method in red and the implicit method in green. In this example, for a relative big timestep $h = 1$ the explicit method does not converge to the exact solution and became unstable while the implicit method does not converge to the exact solution either but the related curve remains under the curve of the exact solution.

$$
\begin{aligned}
\mathbf{F}_{d,\mu}(\mathbf{u}^{i+1}, \mathbf{v}^{i+1}, t_{i+1}) =&\, \mathbf{F}(\mathbf{u}^i, \mathbf{v}^i, t_i) \\
&+ \frac{\partial}{\partial \mathbf{u}}\mathbf{F}(\mathbf{u}^i, \mathbf{v}^i, t_i)(\mathbf{u}^{i+1} - \mathbf{u}^i) \\
&+ \frac{\partial}{\partial \mathbf{v}}\mathbf{F}(\mathbf{u}^i, \mathbf{v}^i, t_i)(\mathbf{v}^{i+1} - \mathbf{v}^i) \\
&+ \frac{\partial}{\partial t}\mathbf{F}(\mathbf{u}^i, \mathbf{v}^i, t_i)h
\end{aligned}
\tag{2.145}
$$

Then by substituting in the Eq. (2.143), then using Eqs. (2.144) and (2.142), and finally arranging for $\Delta\mathbf{v}$ a linear system of the form

$$
\mathbf{A}\Delta\mathbf{v} = \mathbf{b}
\tag{2.146}
$$

is obtained, where $\mathbf{A}$ is the $3n \times 3n$ system matrix, and $\mathbf{b}$ is the $3n$ right-hand side vector:

$$\mathbf{A} = 1 - h\mathbf{M}^{-1}\frac{\partial}{\partial \mathbf{v}}\mathbf{F}(\mathbf{u}^i, \mathbf{v}^i, t_i) - h^2\mathbf{M}^{-1}\frac{\partial}{\partial \mathbf{u}}\mathbf{F}(\mathbf{u}^i, \mathbf{v}^i, t_i) \tag{2.147}$$

$$\mathbf{b} = h\mathbf{M}^{-1}\mathbf{F}_a(t_{i+1}) + h\mathbf{M}^{-1}\mathbf{F}(\mathbf{u}^i, \mathbf{v}^i, t_i)$$
$$+ h^2\mathbf{M}^{-1}\frac{\partial}{\partial \mathbf{u}}\mathbf{F}(\mathbf{u}^i, \mathbf{v}^i, t_i)\mathbf{v}^i + h^2\mathbf{M}^{-1}\frac{\partial}{\partial t}\mathbf{F}(\mathbf{u}^i, \mathbf{v}^i, t_i) \tag{2.148}$$

To assemble $\mathbf{A}$ and $\mathbf{b}$, the Jacobian matrices $\frac{\partial}{\partial \mathbf{u}}\mathbf{F}$ and $\frac{\partial}{\partial \mathbf{v}}\mathbf{F}$ have to be determined. By acknowledging that h^2 is a very small value, the term $h^2\mathbf{M}^{-1}\frac{\partial}{\partial t}\mathbf{F}(\mathbf{u}^i, \mathbf{v}^i, t_i)$ can be neglected for simplification.

To obtain the Jacobian matrix of the deformation forces $\mathbf{F}_d$, for each particle p_l, the deformation forces at current timestep $\mathbf{F}_d(\mathbf{u})$ are calculated. Then a small but finite value δu is added to each element $\mathbf{u}_m$ of the vector $\mathbf{u}$, one element at a time, that results in the varied configurations $\hat{\mathbf{u}}(m)$. Eventually the varied passive forces $\hat{\mathbf{F}}_d(\hat{\mathbf{u}}(m))$ of all particles p_l are calculated. The Jacobian matrix $\mathbf{J}_u$ is then built by calculating the difference quotients:

$$J_{u,m} = \frac{\mathbf{F}_d - \hat{\mathbf{F}}_d(\hat{\mathbf{u}}(m))}{\delta u} \tag{2.149}$$

In a similar way, the Jacobian matrix $\mathbf{J}_v$ of the damping forces $\mathbf{F}_\mu$ is calculated by adding small but finite values δv to each of the element of the velocity vector $\mathbf{v}$ and repeating the steps detailed above. Both Jacobians are then replaced in Eqs. (2.147) and (2.148).

Since any particle of the system is only connected to a maximal number of 26 neighboring particles (voxel's vertices in 3D space) the matrix $\mathbf{A}$ is sparse and it can be shown that every row of the matrix contains a constant count of non zero elements. That allows the use of efficient methods of computational complexity $O(n)$ to solve the linear system (2.146). An iterative solver (GMRES) that takes advantage of the matrix $\mathbf{A}$ sparsity provided with the PETSc package [59] is used.

To take advantage of the possibility of taking bigger time steps which the implicit integration provides without risking the algorithm to becomes unstable, an adaptive time-stepping mechanics is used.

It was observed that when the system shows high dynamics, caused by large forces for example, the number of iterations N_s the iterative solver needs to converge when solving the linear system in Eq. (2.146) increases. If the system shows monotonic behavior, the number of iterations decreases. Furthermore, the number of iterations increases massively just a few steps before the system becomes unstable.

In general the number of iterations appeared to be a good parameter to control the timestep size used in the implicit integration algorithm as shown in Figure 2.21.

By defining a lower and an upper limit for the number of iterations, the implicit solver can adapt the timestep size. If the number of iterations reaches the lower limit, the timestep size is doubled. If the number of iterations reaches the upper limit the timestep size is halved.

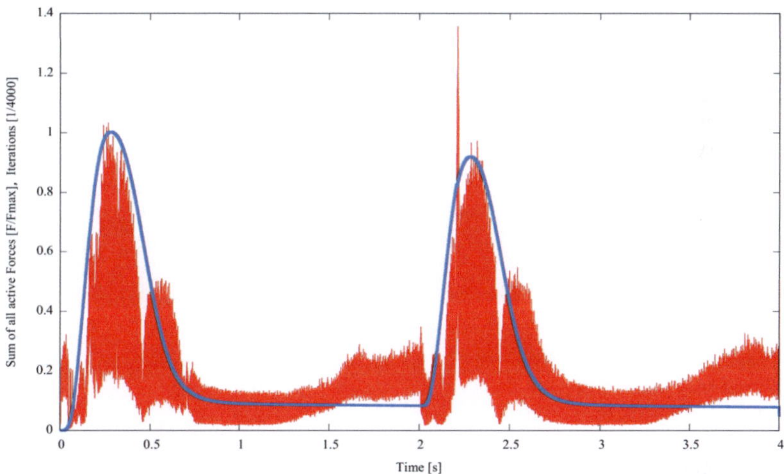

Fig. 2.21 A good correlation between the normalized active forces sum (blue) and the iteration counts of the GMRES iterative solver of the PETSc package (red).

This technique is not completely legitimate, because the mathematics behind it was not carefully analyzed. However, so far experiments are showing this method is able to manage the timestep size with great success.

2.5.6 Critical Timestep Size

To obtain a realistic simulation, the maximum timestep must be smaller than the time, a mechanical wave needs to pass the modeled object. Otherwise the propagation of a mechanical wave cannot be simulated properly. The critical timestep size is given by

$$h_\mathrm{c} = \frac{L}{c} \tag{2.150}$$

where L is the length of the smallest side of the volume element and c is the propagation velocity of a mechanical wave (the speed of sound). The propagation velocity depends on the damping behavior of the object where the wave is propagating. An appropriate timestep size can be obtained, by comparing the results of the simulation for different timestep sizes. A timestep is sufficiently small, if the results of a simulation using that timestep and the results of a simulation using a timestep which is orders of magnitude smaller do not differ substantially. This method was used to justify the timestep used for a given application.

2.6 Boundary Conditions

To constrain the movement of the modeled object, particles can be marked with different fixation and control tags. A file containing the indices of the particles and the associated tags is processed during the initialization of the system to mark the particles.

Fixation tags can be one of three possible tags and their combination. These tags are:

- Fixed in $x-$axis direction
- Fixed in $y-$axis direction
- Fixed in $z-$axis direction

Here, the axes correspond to the global coordinate system.

For example, a particle p_i marked with the $x-$axis and the $y-$axis fixation tags, can only move in the $z-$axis direction. In the case p_i was marked with all fixation tags, it will have the same position during the entire simulation.

The code implementation of the fixation boundary condition involves iterating over particles and setting the offset of the particle according to the fixation tags to zero. However in the case when the implicit time integration is used, the x,y,z mass components of the mass tensor $\mathbf{M}$ are set to a very large value according to the fixation tags, preventing it from moving.

Control tags can be one of three possible tags and their combination. These tags are:

- Controlled in $x-$axis direction
- Controlled in $y-$axis direction
- Controlled in $z-$axis direction

For example, a particle p_i marked with the $x-$axis control tag, will be moved every simulation step with a predefined step in the direction of the $x-$axis.

Using both fixation and control tags gives the possibility of conducting strain-stress simulations. For example a uni-axial strain-stress experiment of a patch of elastic material can be conducted by fixing the model from both ends in all directions, and controlling one end in the direction of one of the axis.

2.7 System's Output

To visualize and evaluate the deformation, the system exports periodically sets of data related to the simulation. The export-period can be set at the beginning of the simulations depending on the simulation's circumstances.

The system can export the coordinates, offsets, velocity and acceleration vectors of all particles. Other general data, not related to individual particles, can also be exported. These comprises the simulation time step, the particles total kinetic energy, the system's barycenter coordinates, the system's total volume.

Beside plain text data files, the system supports several file formats like the well known Visualization Toolkit (VTK) file format [60].

2.8 System Verification

In this section, several simulations designed to test the functionality of the system and verify its implementation are demonstrated.

2.8.1 Reproducing Mechanical Properties

To verify the ability of the system to reproduce the mechanical properties of a modeled material represented by its constitutive law, a series of uniaxial stretch and simple shear simulations were conducted and the resulting stress strain relations were compared with the curves calculated theoretically. In cases where the theoretical calculation of the stress strain relation is not trivial, like in simple shear experiments, the applied work to the system is compared with the potential deformation energy. If the stresses were calculated correctly, the potential deformation energy must be equal to the applied work. The deformation energy can be calculated easily using the deformation tensor and the energy density function. The linear approximation of work W_{i+1} applied between stretch steps s_i and s_{i+1} is given by

$$W_{i+1} = \sum_{n=1}^{N} \left(\mathbf{x}_n(s_{i+1}) - \mathbf{x}_n(s_i) \right) \cdot \frac{\mathbf{f}_n(s_{i+1}) + \mathbf{f}_n(s_i)}{2} \tag{2.151}$$

where $\mathbf{x}_n$ are the coordinates of the controlled point p_n, $\mathbf{f}_n$ is the force acting on p_n, and N is the number of controlled particles. The total work is given by

$$W_{\text{total}} = \sum_{i=1}^{i_{\text{total}}} W_i \tag{2.152}$$

For the simulations, the constitutive law of myocardial tissue proposed by Hunter *et al.* [61] was used:

$$W = k_{11} \frac{E_{11}^2}{|a_{11} - E_{11}|^{b_{11}}} + k_{22} \frac{E_{22}^2}{|a_{22} - E_{22}|^{b_{22}}} + k_{33} \frac{E_{33}^2}{|a_{33} - E_{33}|^{b_{33}}} \tag{2.153}$$
$$+ k_{12} \frac{E_{12}^2}{|a_{12} - E_{12}|^{b_{12}}} + k_{13} \frac{E_{13}^2}{|a_{13} - E_{13}|^{b_{13}}} + k_{23} \frac{E_{23}^2}{|a_{23} - E_{23}|^{b_{23}}}$$

where E_{ij} are the elements of the Green strain tensor $\mathbf{E}$. They are related to the deformation tensor $\mathbf{F}$ according to Eq. 1.6. k_{ij} and b_{ij} are parameters obtained by experimental data. The parameters used in this simulation are given in Table 2.1. For more about the implementation of this constitutive law see Section 4.7.5.

Table 2.1 Parameters for the energy density function proposed by Hunter *et al.* adapted to canine myocardium (Parameters from [61, 23])

k_{11}	k_{22}	k_{33}	k_{12}	k_{13}	k_{23}
2.842	0.063	0.31	1.0	1.0	1.0

a_{11}	a_{22}	a_{33}	a_{12}	a_{13}	a_{23}
0.318	0.429	1.037	0.731	0.731	0.886

β_{11}	β_{22}	β_{33}	β_{12}	β_{13}	β_{23}
0.624	2.48	0.398	2.0	2.0	2.0

It is easy to recognize from Eq. (2.153) that this material model has three directions of passive mechanical anisotropy, along the material local coordinate system that represents the fiber, sheet and sheet-normal directions of myocardial tissue, which we will call the fiber coordinate system.

First deformation forces calculation using axial and torsion springs is put to test. Polynomial axial springs parameters were obtained by fitting a third order function to the strain-stress curves of Eq. (2.153) for the uniaxial stretch experiments (see Section 2.4.2.1). For torsion springs, the values were chosen to give a good approximation of the shear curves (see Section 2.4.2.1). The parameters are listed in Table 2.2, the values were fitted to maximum strain of 20%.

Table 2.2 Parameters (kPa) of third-order polynomial axial and torsion springs. The third-order polynomial functions were fitted to the constitutive law of Hunter *et al.* [61]

k_{11}	k_{13}	k_{21}	k_{23}	k_{31}	k_{33}
6.358	768	0	288.525	0.305	5.699

$k_{12,1}$	$k_{12,3}$	$k_{13,1}$	$k_{13,3}$	$k_{23,1}$	$k_{23,3}$
3.355	97.374	97.374	3.355	45.896	2.261

Uniaxial stretch experiments of a patch of $6 \times 6 \times 10$ hexahedra were conducted. The model's resolution was 5mm and the patch was divided to 2160 tetrahedra using a tetrahedral mesh topology. Forces were calculated using the axial and torsion springs of the third order and the parameters in Table 2.2. The resulting stress strain curves are plotted in Figure 2.22 alongside the theoretical calculation of stress strain relation of Eq. (2.153). It is important to remember that the parameters are related to the stress values in volume elements and not to forces, therefore these curves show that the behavior of the models is independent of the chosen resolution. However, a better behavior is generally expected at higher resolutions specially in case of modeling objects with smooth or sharp edges. Then the method of virtual hexahedron was put here test. Uniaxial stretch simulations of a patch of myocardial tissue along the fiber, sheet and sheet-normal axes were conducted. Hereby, stretching is performed with small steps s_i, and the resulting stress at each stretch step is determined after the system reaches an equilibrium state. In all simulations, deformation tensors are calculated using the shape functions method (see Section 2.4.2.2).

First stretch simulation of a hexahedron, using tetrahedral mesh topology where each hexahedron was divided to six tetrahedra, were conducted (Fig. 2.23). Using the same mesh topology, simulations of a cuboid composed of $4 \times 4 \times 8$ hexahedra (768 tetrahedra) were performed (Fig. 2.25). In all these simulations, the fiber coordinate system was set parallel to the regular grid. To show that the system can control anisotropy, the fiber coordinates system was rotated by $\phi = 21.25°$ and $\Theta = 49.6°$ in spherical coordinates, and stretch simulations of the $4 \times 4 \times 8$ cuboid model in the global $z-$axis direction were conducted (Fig. 2.25).

In all simulations, the simulation results are in very good agreement with the theoretical calculations. Other simulations not presented here using the hexahedral topology gave results equal to the results for tetrahedral mesh topologies. Simple shear experiments were conducted, to validate the ability of the system to reproduce shear forces. In order to do that, the top plane of the modeled patch was displaced in a direction parallel to the plane. For each experiment, three different shear procedures were simulated: The plane orthogonal to the fiber axis was displaced in direction of the sheet-normal axis, the plane orthogonal to the sheet axis was displaced in direction of the fiber axis and the plane orthogonal to the sheet-normal axis was displaced in direction of the sheet axis. Models used for the

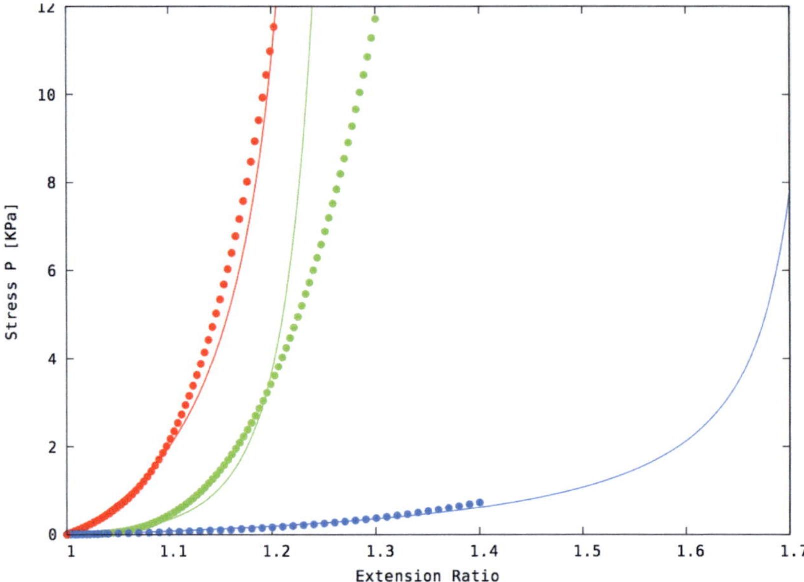

Fig. 2.22 The stress-strain relation of an 6x6x10 hexahedra object of 5 mm resolution (2160 tetrahedra), for strains in fiber (red), sheet (green) and sheet-normal (blue) directions. Forces were calculated using the axial and torsion springs of the third order. The results of the simulations (circles) using third degree spring functions were used are displayed with theoretical calculation (solid lines). The object was stretched to 20% of its original length.

stretch simulations were used for these experiments. However no fiber coordinates rotation was applied.

To validate the simulation results, in addition the applied work to the system was compared with the deformation energy.

Figures 2.26 and 2.27 show the simulations' outcome. Here the difference between the multiple hexahedra cuboid and the single hexahedron deformation energies is due to the contribution of the axial strain to the deformation energy in the case of the cuboid, resulting in bigger differences between the three curves in comparison with the single hexahedron case. Nonetheless, the resulting applied work in both simulations fits very well to the deformation energy.

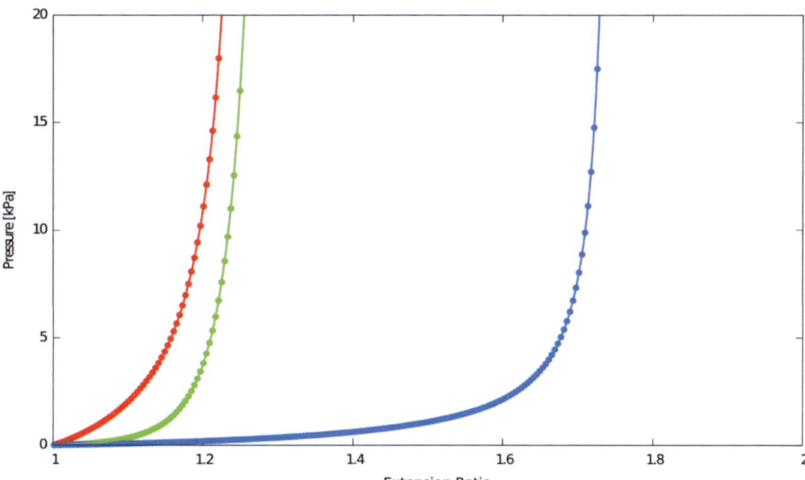

Fig. 2.23 Stress-strain curves for uniaxial stretch of a voxel (6 tetrahedra) along all axes. Simulation output (points), theoretical calculation (solid line) for fiber (red), sheet (green) and sheet-normal (blue) directions.

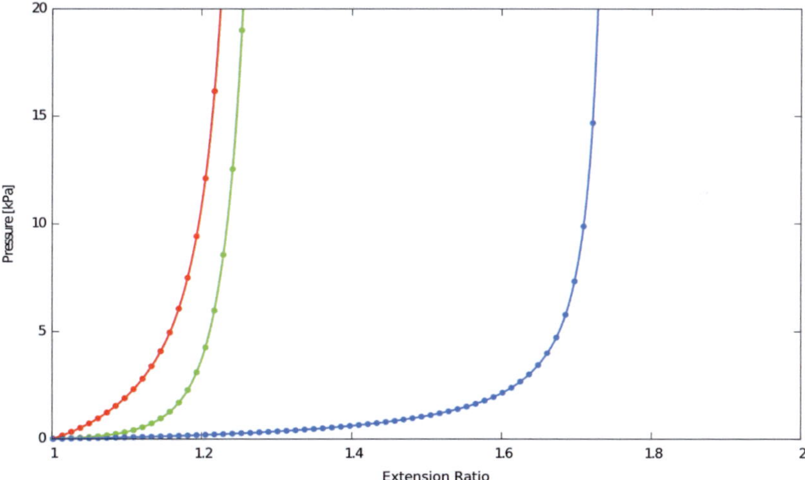

Fig. 2.24 Stress-strain curves for uniaxial stretch of a 4x4x8 voxel object (768 tetrahedra) along all axes. Simulation output (points), theoretical calculation (solid line) for fiber (red), sheet (green) and sheet-normal (blue) directions.

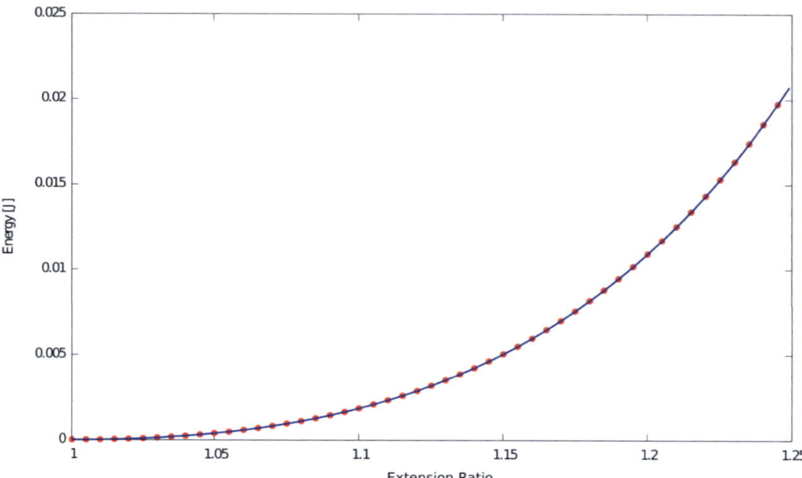

Fig. 2.25 Uniaxial stretch along z-axis, fiber coordinate system is rotated relative to the hexahedral mesh with $\theta = 49.6°$ and $\phi = 21.75°$. Simulation output or applied work (points), theoretical calculation or deformation energy (solid line), 4x4x8 voxel, 768 tetrahedra.

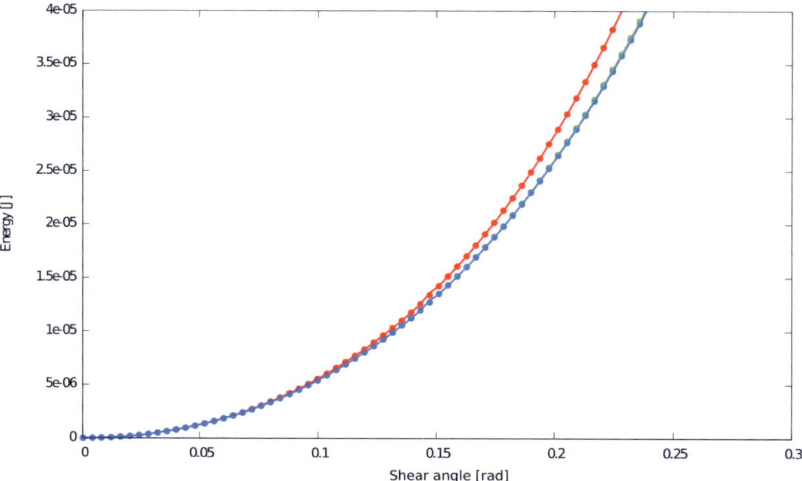

Fig. 2.26 Energy-shear curves for simple shear. In the legend the first axis expresses the normal vector of the displaced plane and the second axis expresses the direction of displacement. Simulation output or applied work (points), theoretical calculation or deformation energy (solid line) in fiber (red), sheet (green) and sheet-normal (blue) directions for 1 voxel, 6 tetrahedra.

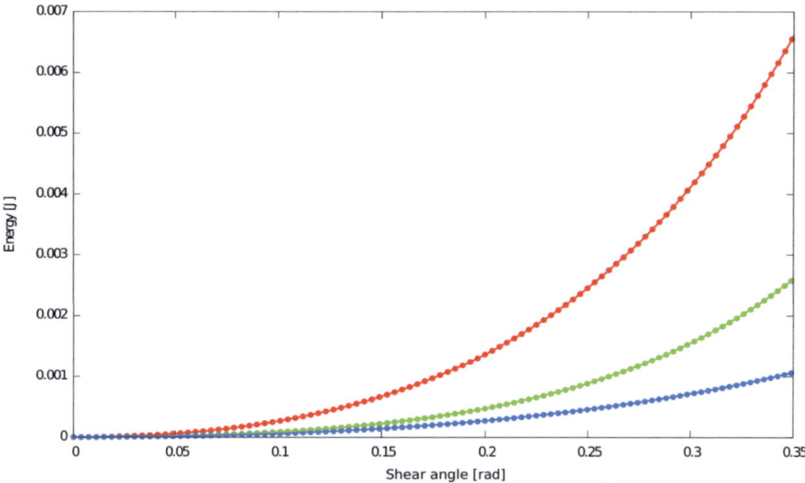

Fig. 2.27 Energy-shear curves for simple shear. In the legend the first axis expresses the normal vector of the displaced plane and the second axis expresses the direction of displacement. Simulation output or applied work (points), theoretical calculation or deformation energy (solid line) in fiber (red), sheet (green) and sheet-normal (blue) directions for 4x4x8 voxel, 768 tetrahedra.

Many other validation simulations using different mesh topologies and both methods to calculate the deformation tensor, and different geometries were conducted. In general as long as no hour-glassing occurs (see Section 2.9.1 and Fig. 2.34), the hexahedral mesh topologies can reproduce the mechanical properties of the material almost as good as the tetrahedral mesh topologies. A small difference can be found between the results when hexahedral, or tetrahedral volume elements are used. This discrepancy results from the linear interpolation of the deformation tensors. These simulations show as well that the system is able to calculate the passive forces of myocardial tissue correctly, and to correctly reflect anisotropy using the method of virtual hexahedron, that facilitates to a great extent modeling different deformable materials by using their constitutive laws.

2.8.2 Volume Preservation

To demonstrate the ability of the presented mass-spring system to maintain its volume under deformation, the deformation of a cuboid of $6 \times 6 \times 12$ hexahedral elements under gravity loading was simulated using implicit time integration, and the continuum mechanics method for volume preservation (see Section 2.4.2.3). In the presented simulations, different values for the volume preservation coefficient p were used, and for each simulation the relative change of volume $\frac{\Delta V}{V^0}$ was measured.

Figure 2.28 shows $\frac{\Delta V}{V^0}$ as a function of time for different values of p. For this application a value of p bigger than $10^3 kPa$ will keep the relative change of volume well below 1%.

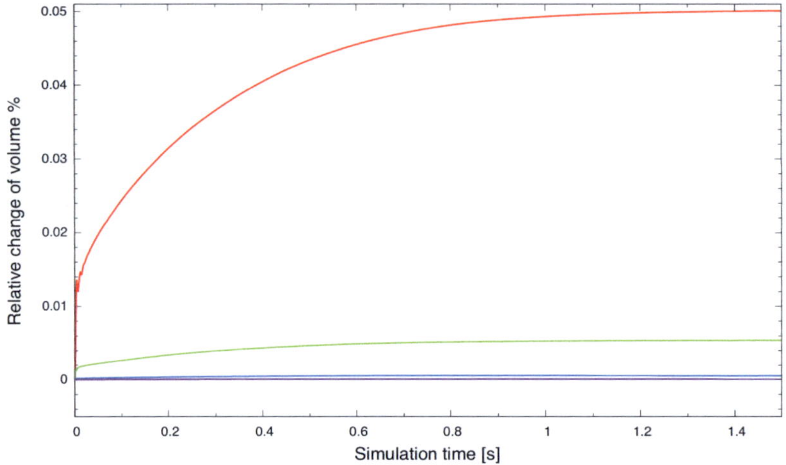

Fig. 2.28 The relative volume change curve during gravity loading simulations with coefficients: $p = 10^2 kPa$ (red), $p = 10^3 kPa$ (green), $p = 10^4 kPa$ (blue) and $p = 10^5 kPa$ (magenta).

2.8.3 Computational Complexity

The main advantage of using mass-spring system in physical based modeling of deformable objects is that each time step iteration has a computational complexity of $O(n)$ where n is the number system's particles. The reason is that determining the deformation in each time step requires solving the system of ODEs given in Eqs. (2.125) and (2.126) which can be done in a single sweep over all system's particles using explicit time integration methods.

Implicit time integration requires solving a linear system of the form given in Eq. (2.146) where the system matrix $\mathbf{A}$ is $3n \times 3n$ dimensions. Solving such systems is a task of $O(n^2)$ complexity. However, as mentioned in Section 2.5.5, the system matrix is sparse and therefore efficient methods of $O(n)$ complexity that take advantages of the sparsity of the matrix can be used to solve the system.

To demonstrate the computational complexity of the system in the case of using implicit time integration, cuboid-shaped objects of dimensions $6 \times 6 \times 12\ cm$, sharing the same physical properties but having different mesh resolutions and thus different number of elements, were simulated using the same simulation setup, and the same computational environment. In each simulation the computation time was measured every 1000 timestep iterations. Hereby, a fixed time step was used.

Figure 2.29 shows the average computation time of 1000 iterations for the corresponding volume elements count. For simulations repeated more than once, error bars showing the deviation from the average are shown in the figure. The results show clearly the linear relation between computation time and the number of volume elements simulations with the implemented implicit time integration method.

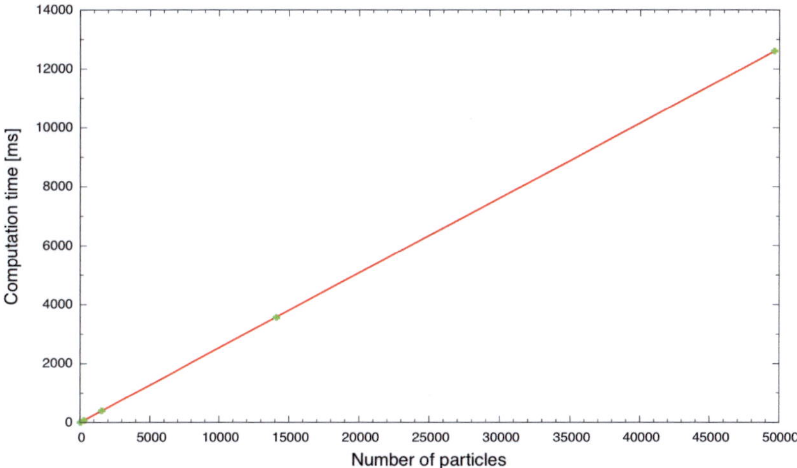

Fig. 2.29 The average computation time needed for 1000 timestep iterations with a constant timestep. The measured average and the standard deviation (green) and the linear regression (red).

2.9 Discussion

The system is a combination of several modules. In each module, there are many design aspects that affect the implementation and the simulation outcome of the system. Also for each modeling task or application, there are different modules combination that could be used. Choosing between these combinations affects the system in many areas like the stability of the simulation, accuracy of simulation results, computation duration ... etc. In the following, we will go over the modules while discussing the most important of these issues.

2.9.1 About mesh topologies

Each of the mesh topologies implemented in this system has advantages and drawbacks.

For instance, the greatest advantage of using a hexahedral mesh topology is that most medical imaging devices, like CT and MRI, produce images stored in a three dimensional lattice. That makes models generation of objects depicted in medical images a straight-forward task. Nonetheless, in order to model smooth surfaces using hexahedral mesh topology, a high resolution mesh must be used. That means more elements must be modeled leading to the increase of computational costs.

As mentioned in 2.1.1, the deformation of the model is represented by the offset of the vertices from their initial positions. Since the time integration module updates the particles' coordinates regardless of volume elements they belong to, it is possible that a particle of a hexahedron will move across one of that hexahedron's faces, turning it to a non-simple polyhedron as shown in Figure 2.30. Such elements spoil the calculation of inner forces that depends on the volume elements integrity (see Section 2.4.2) therefore they are called corrupt. Corrupt elements

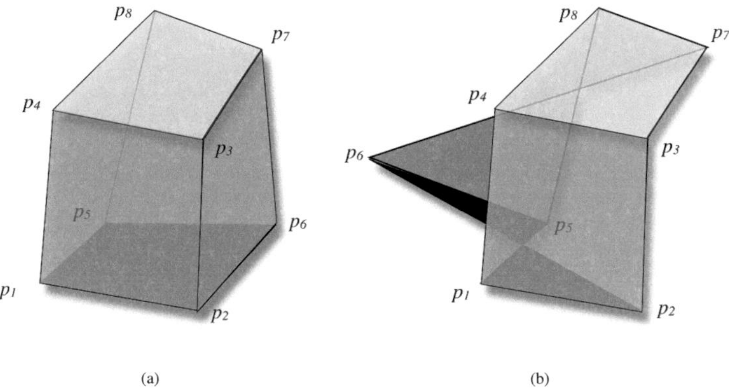

(a)   (b)

Fig. 2.30 A deformed hexahedrom (a) and a non-simple polyhedron where vertex p_6 of the hexahedron moved across the opposite face of the hexahedron defined by p_1, p_5, p_8 and p_4 (b).

can be either the result of numerical instability or the result of using a timestep for time integration bigger than the critical timestep h_c (see Section 2.5.6). The modeling framework must implement methods to detect these elements and eventually to deal with them. Implementing a method to determine whether during deformation a hexahedron became corrupt is not trivial.

Faces of eight nodes hexahedron (linear hexahedron) are under-determined, because each of the faces is defined by four points. Since the vertices coordinates determine the deformation, there is no way to tell whether a linear hexahedron is in a state shown in Figure 2.31(a) or in state shown in Figure 2.31(b) or in some other state.

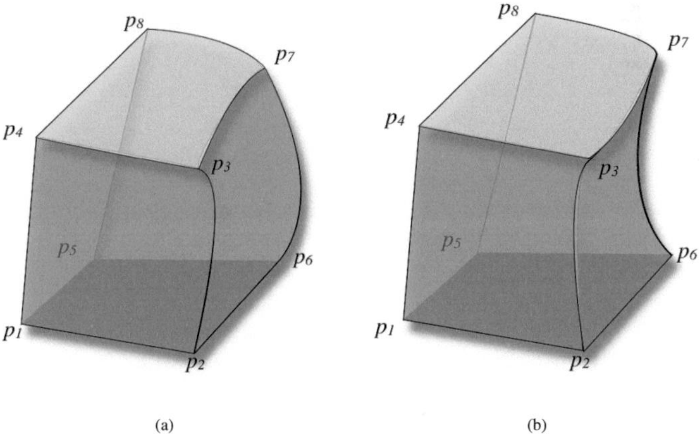

Fig. 2.31 Two deformed hexahedral elements sharing the exact same vertices position, the first with concave surfaces (a) and the other with convex surfaces (b).

Determining the surfaces of volume elements' faces, the normals on the faces, and the volumes of the elements are routine tasks essential for the calculation of inner forces (see Section 2.4.2). Every time the surface of a deformed linear hexahedron's face, the normal to that face, and the volume of the hexahedron are computed, a systematic error is made.

Interpolating the coordinates of the intersection points is also prone to this systematic error.

To mitigate the errors in calculating the surface of faces defined by four points, the surface is divided to two triangles in two different ways as shown in Figure

2.32, and the surfaces of the resulting triangles are averaged. Similarly the normal to the face is determined by calculating the average of normals of the four mentioned triangles.

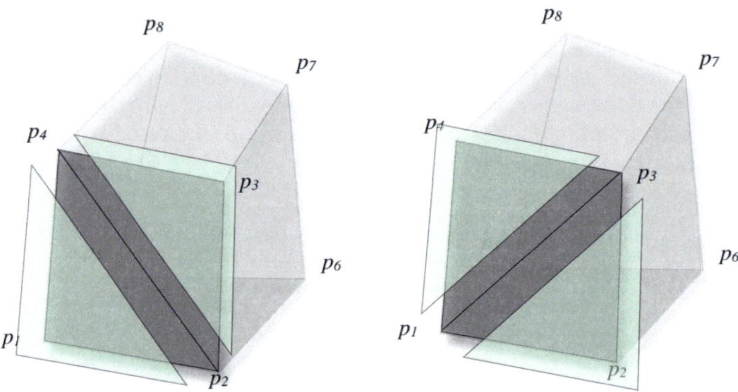

Fig. 2.32 Two different ways to divide a face of a hexahedron in two triangles.

The volume of each hexahedron is computed using a method based on dividing the hexahedron to five or six tetrahedra as in Figures 2.7 and 2.8, and then summing up the computed volumes of the resulting tetrahedra as in in the work of Grandy *et al.* [50].

Furthermore, the deformation state of a hexahedron is not totally defined by the positions of its intersection points as displayed in figure 2.33. Therefore the forces resulting due to deformation in both cases are equivalent, which does not reflect the physical case.

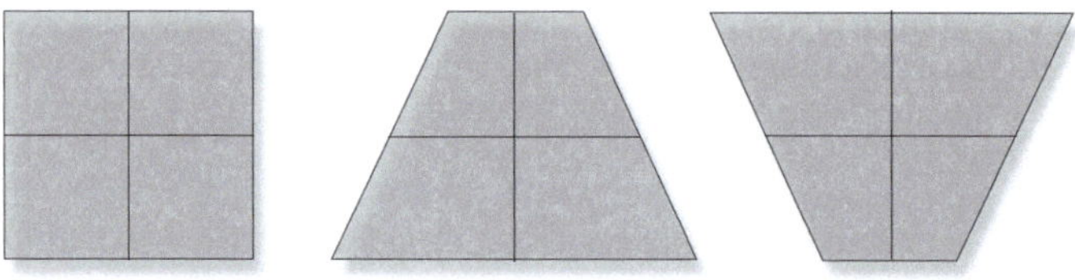

Fig. 2.33 One face of three different deformed hexahedra with equal intersection points configuration showing that hexahedra are not uniquely defined by the intersection points.

On top of all that, hexahedra suffer from hourglass modes which are non-physical zero deformation energy modes that can result in non-physical deformations [1]. Figure 2.34 shows the different hourglass modes of linear hexahedron models.

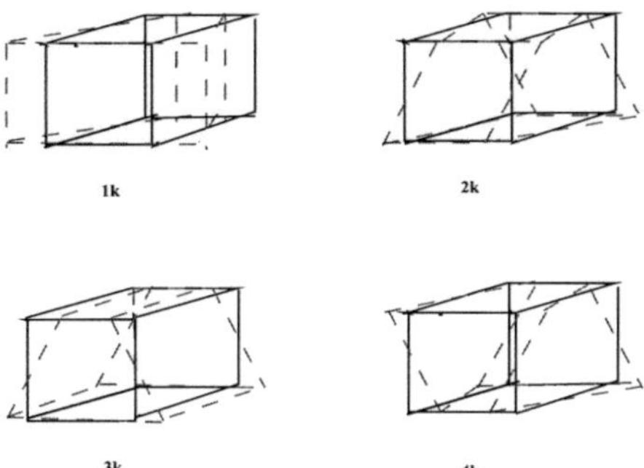

Fig. 2.34 Hourglass modes of the linear hexahedron from [62].

For all these reasons several tetrahedral mesh topologies were examined and implemented. Most of the simulations conducted in this work are based on tetrahedral mesh topologies.

Using tetrahedra instead of hexahedra is very beneficial from the implementation and the simulation outcome points of view.

The triangle faces of tetrahedra are always uniquely defined, hence surfaces and normals on surfaces can be computed with no systematic error using Eqs. (2.14) and (2.11). For the same reason, coordinates of intersection points can be calculated uniquely using the vertices of the faces to which the points belong. Reversely, the deformation of a tetrahedron can be uniquely retrieved by the intersection points and the interpolation coefficients related to these intersection points.

A deformed tetrahedron is uniquely defined by its vertices, therefore the volume V_k of a tetrahedron can always be calculated with

$$V_h = (\mathbf{x}_1 - \mathbf{x}_4) \cdot \left((\mathbf{x}_2 - \mathbf{x}_4) \times (\mathbf{x}_3 - \mathbf{x}_4) \right) \tag{2.154}$$

$$V_k = \frac{1}{6} |V_h| \tag{2.155}$$

where V_h is the signed volume of the parallelepiped defined by the four tetrahedron vertices $p_i (i = 1, \ldots, 4)$.

V_h can be used to detect corrupt tetrahedra. A corrupt tetrahedron is detected if the value of V_h changes sign during deformation. In the case all vertices of a tetrahedron become coplanar, many equations to calculate inner forces become invalid. Therefore flat tetrahedra are equally unpleasant as a corrupt tetrahedron and must also be detected.

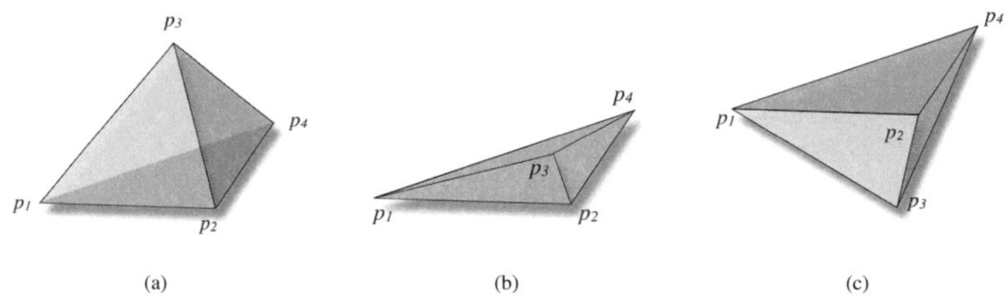

Fig. 2.35 A healthy tetrahedron (a), a tetrahedron with coplanar vertices (b) and a corrupt tetrahedron (c).

Tetrahedra do not have hourglass modes and therefore do not suffer from that phenomenon. But they do suffer from another phenomenon called locking that occurs when a hard constraint on volume preservation is imposed [49]. Locking manifests itself as a catastrophic artificial stiffening of the system. One practical solution to reduce the effect of locking is not to use fully but nearly incompressible material laws. The more the volume is allowed to change the less the effect of locking is observed.

The tetrahedral mesh topologies based on regular grids inherit some of the drawbacks of hexahedral mesh topologies. For instance modeling smooth surfaces needs high resolution meshes. Furthermore, the number of tetrahedra needed to model an object in comparison with a hexahedral mesh topology with the same resolution, is five or six times greater than the number of hexahedra, increasing the computation cost accordingly.

Tetrahedral mesh topology based on unstructured grids, have many advantages over the other implemented topologies. By using a good mesh generator and the right set of mesh optimization and filtering, it is possible to obtain meshes that have smooth surfaces starting from 3D lattice image datasets. That eliminates the need of high resolution models in order to model smooth surfaces needed in topologies based on regular grids.

But as it is often the case, nothing is for free. Here other problems arise, like volumetric locking and inhomogeneous distribution of volume. These problems must be addressed in order to mitigate their undesirable effects. To reduce the effect of inhomogeneous distribution of masses, a maximum limit is imposed on tetrahedron volume during the mesh generation phase, at the same time increasing the number of tetrahedra needed for an object and consequently the related memory and computational cost.

Image information relevant to modeling, for example the anisotropies distribution, are usually available in the form of a lattice dataset related to the original lattice image dataset. Assigning this information to the volume elements when using a hexahedral mesh or a tetrahedral mesh topology based on regular grid is straight-forward, but in the case of unstructured grids, this information must be set for generated tetrahedra that could extend over one or more elements with different values of the information lattice. One possible method to set these values for the generated tetrahedra is using interpolation techniques. In this work linear interpolation was used. However better interpolation techniques could be implemented.

Tetrahedral mesh generating methods which do not impose a limit on resulting tetrahedra's minimal volume, may generate some very skinny tetrahedra. Such tetrahedra are called *slivers*. Slivers are more prone to becoming corrupt or flat because of their morphology. A sliver removal algorithm or a better mesh generator can be used to avoid problems associated with these tetrahedra.

2.9.2 About Forces Calculation

Several issues concerning the calculation of deformation, active and volume preservation forces are mentioned here. Starting with deformation forces, sometimes material's constitutive laws are given only for the stretch direction. These constitutive laws do not provide information about stress in case of compression. If internal deformation forces are set to zero when modeling the compression of such materials, and no other forces can work against the compression like volume preservation forces, the elements can collapse and become corrupt making the results of the simulation invalid. A method to maintain a minimal volume for volume elements that can sustain compression must be implemented.

Some material's constitutive laws contain a pole in their formulas. When strain approaches a pole, stress rises immensely dividing the strain axis to two ranges. The range of valid strains and the range of invalid strains. Because the timestep used for time integration is not infinitesimally small, a volume element might deform from a state where the element's strain is within a valid strain range to a state where the element's strain is invalid. In that case the constitutive law formula might deliver a stress, and the simulation continues, but since the strain is invalid that stress has no physical meaning, and the simulations results become invalid. Therefore constitutive laws that contain a pole are substituted with a high-order polynomial approximation which is inherently pole-free.

Volume preservation forces based on continuum mechanics solution according to Eq. (2.86) uses the parameter p related to the bulk modulus. The parameter p has to be chosen in a way such that the volume preservation forces resulting from a very small change in volume, have the same magnitude as all the forces participating in the deformation. When the volume deviates from its original value, hydrostatic work is added to the system and eventually, giving rise to penalty forces that trys to keep the change in volume around zero.

Locking associated with tetrahedral mesh topologies can be avoided by modeling the absolutely incompressible material with a nearly incompressible model. This can be done by choosing a value for p that is not too high allowing the volume to vary, but so small that the volume variation won't exceed an accuracy limit. If a suitable value of p cannot be found for a certain application, tetrahedral mesh topologies should be avoided, or different volume preservation force models must be implemented.

Material specific friction detailed in Section 2.4.2.5 was implemented only for axial friction or damping and no shear friction was implemented. If needed, this could be done.

The model used for active forces sets the forces or tension to the main anisotropy direction. To simulate internal processes producing forces with components in all anisotropy direction, a different active forces model must be developed and implemented.

2.9.3 About Time Integration

When using the forward Euler method or the prediction-correction method, the volume of the system V oscillates around the original volume of the system V^0.

An explicit solver considers only the present forces to calculate the solution for the next timestep. Upcoming volume preservation forces caused by a change in volume, are not regarded. Since small changes in volume produce huge volume preservation forces, tiny timesteps must be taken, to avoid instability. However, even for small timesteps the volume preservation forces can cause a strong undesirable oscillation of the volume.

To get around these problems the upcoming volume preservation forces of the next timestep have to be taken into account, for solving the equations of motions.

Therefore the implicit backward Euler time integration method is favorable. Hereby, the solution of the next timestep depends on the forces which will arise at that step. This method ensures volume preservation and reduces the mentioned volume oscillation even for larger timesteps.

Implementing the implicit time integration for solving the equations of dynamics not only increased the stability and allowed for larger timesteps, but also removed the unwanted oscillation of the volume mentioned above. Implicit time integration introduces artificial damping that contaminates the simulation results. The larger the timestep size is the more artificial damping is added to the system. Therefore the timestep should be chosen carefully in order to obtain valid simulation results.

In the implementation of the implicit integration method two approximations are made. The first is making the first order approximation of the Taylor series expansion of $\mathbf{F}(\mathbf{u}^{i+1}, \mathbf{v}^{i+1}, t_{i+1})$ in Eq. (2.145), and the second is neglecting the term $h^2 \mathbf{M}^{-1} \frac{\partial \mathbf{F}}{\partial t}(\mathbf{u}^i, \mathbf{v}^i, t_i)$ in the same equation. Because of these approximations the stability of the implemented implicit time integration method is not unconditional. That means large changes in forces can cause numerical instability even when the implemented implicit time integration method is used for the integration.

Friction in general affects the speed of mechanical waves propagation. According to Eq. (2.150) the critical timestep h_c depends on the mechanical wave speed. It must be kept in mind that changing friction conditions might require a change in timestep used for time integration.

Chapter 3
Modeling Breast Elastomechanics with Adamss

3.1 Motivation

According to the World Health Organization (WHO), breast cancer is the most common cancer in women. It represented about 14% of all female cancer mortalities in 2008 world-wide [63]. Although the incidences rate in the developed world is still higher, it is increasing in the developing world due to increased life expectancy, urbanization and the adoption of western lifestyles [64].

Early breast cancer detection is the cornerstone of breast cancer control [65]. This can be achieved through regular screening.

Mammography, that involves using a low-dose X-rays to detect characteristic masses or micro-calcifications that are indicators of cancer in the breast, is currently the method most prominent of all screening methods, because it is simple, has relatively low-costs, and clinical trials have proven it helps reducing the mortality rates [66]. Mammography is associated with exposure to ionizing radiation and suffers from high false negative rates ranging from 4% up to 34% [67] and a recall rate of 11% [68]. It is less sensitive in women with radiographically dense breast tissue [69, 70, 71] and requires uncomfortable compression of the breast. These drawbacks, legitimize the search for a safe, low-cost screening method that can replace mammography.

Self breast exam is another screening method, that involves the woman herself looking at and feeling each breast for possible lumps, distortions or swelling, while clinical breast exam is performed by a physician. Beside that self examination found to empower women in many developing countries, giving them the sense of responsibility of their own health, a review concluded that screening by breast self-examination or physical examination does not benefit in reducing mortalities. It rather, inflict harm by resulting in an increased number of benign lesions identified and an increased number of biopsies performed [72]. Hence, self examination is not recommended as a regular screening method.

Breast ultrasound was considered as one of the promising screening tools since its early days. So far, studies have shown that it is less effective than Mammography due to its high false-positive rate and its inability to identify micro-calcification [73]. As new high-resolution, real-time equipments and improved scan techniques are being developed, the role of breast ultrasound is being re-explored [74].

Microwave imaging for breast cancer detection is a potential alternative that is still under investigation and development. The differences in electrical properties of healthy tissues and malignancies give microwave imaging techniques the potential for detecting and monitoring malignant tissues. In recent years, microwave imaging gained more attention due to advances in imaging algorithms, microwave hardware and computational power [75].

For the development of microwave imaging systems, realistic anatomical models comprising detailed dielectric properties have to be used in order to simulate clinical reality. It was found that including the complexity and heterogeneity of breast tissues dielectric properties is very important in clinical microwave studies [76, 77].

Zastrow *et al.* [78] presented anatomically realistic numerical breast phantoms with accurate dielectric properties for a wide range of frequencies generated from MRI datasets of women in prone position.

These phantoms can be used for developing systems in a screening configuration where the patient lies in a prone position. The prone position configurations allows for immersion of the breast in a liquid coupling medium that has a dielectric constant similar to that of the skin what reduces the signals reflection off the skin [79, 80]. Supine position configurations are also under investigation. In these configurations, the patient is examined in supine position without the use of a coupling medium. They are considered less discomforting, as they do not involve immersion of the breast. The supine position causes the breast to flatten on the chest wall, the breast extends in area, and the distance between the nipple and the chest wall decreases to around $5cm$ in average, making the complete volume of the breast within the detection range of currently available radar-based microwave imaging systems [81, 82, 83].

To obtain breast phantoms suitable for developing microwave imaging systems based on the supine position configuration, a method to calculate the deformation of breast phantoms of Zastrow *et al.* from prone to supine position is presented in this chapter. The modified mass-spring system (Adamss) is then used to obtain several of these phantoms in supine position. A short overview of breast anatomy, breast physiology and breast tissue elastomechanics modeling are first presented.

3.2 Breast Anatomy

In this section, a brief overview of the gross anatomy of the female breast based on the work of Sir A. Cooper [84], and of Ramsay *et al.* [85] is presented.

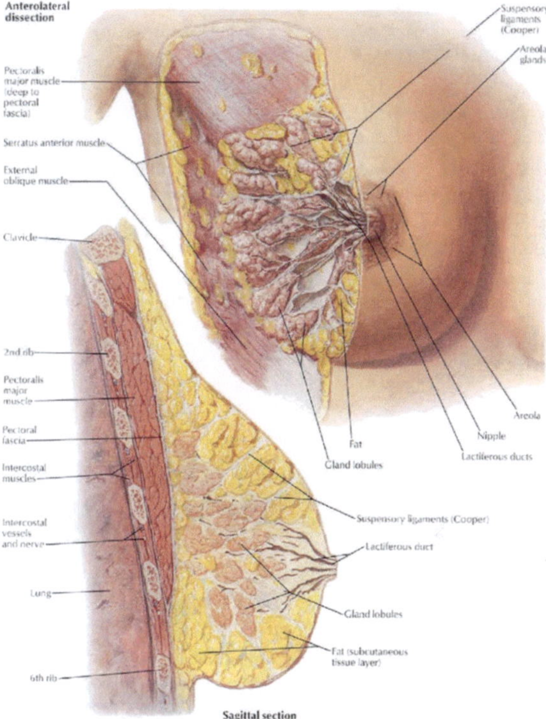

Fig. 3.1 Gross anatomy of the breast (modified from [86]) and edited according to [85].

Figure 3.1 depicts the gross anatomical structure and organization of the female breast. The mammary glands are situated on the anterior and lateral parts of the chest. They extend from the second or third rib, to the sixth or seventh rib. The inner (sternal) side of the breast rests on the pectoralis major and cartilages of the ribs, while the outer (axillary) side rests on the fascia of the thorax, serratus major and external oblique muscles. The breast is very dense in axillary margin and the abdominal margin at the seventh rib, while it is much thinner at the sternal and superior margins at the third rib.

The nipple, an external part of the breast, is a conical shaped, cutaneous projection on the breast, which is attached to the areola at its base. The nipple contains blood vessels, termination of nerves and the ducts from the internal breast glands (carrying milk), all encased by fibrous tissue. It is covered by a texture called the cuticle. The areola is the part of the skin that forms the circular base of the nipple.

The internal structure of the breast is far more interesting. The fascia mammae are divided into a superficial and a deeper layer of the breast, between which the gland of the breast is situated. Both layers are attached to the ligamentous substance that covers the sternum. The superficial layer forms a fibrous cover, passing between the gland and the skin and also entering the interior of the gland. Fibrous extensions called Cooper's ligaments, are sent from the superficial layer to the posterior surface of the skin. Cooper's ligaments provide firmness to the breast and are spread over the top of the secretory gland at the superficial layer. They support folds of the glandular structure and connect the portions of the glands to each other by penetrating the gland. The ligaments also connect the nipple to the glands. Figure 3.1 shows Cooper's ligaments connecting the breast to the skin. The deeper layer of the breast also sends fibers into the gland to unite its parts, and sends other fibers backwards to attach to the pectoralis major.

Fat, the predominant tissue in the breast, fills up the depressions between glandules, large lobes between Cooper's ligaments, between anterior folds of the glandular tissue and between the two layers of fascia mammae. As a result, the breast is cushioned between these two fibrous layers and the fat lobes between the Cooper's ligaments to protect it from pressure and harm.

The breast constitutes a number of glandules connected by the fibrous or fascial tissue of the gland. Each glandule is filled with a large number of elastic milk cells, also known as alveoli, that store the milk that is secreted from the blood carried through arteries. Lactiferous ducts originate from the glandules, they transport the milk from the glandules from which they originate to the apex of the nipple, according to [87, 85] between four to eighteen of these tubes reach the nipple. The network of milk ducts is inhomogeneous, complex and not always arranged symmetrically nor in a radial pattern, as previously described and depicted by [84]. At the apex of the nipple, the lactiferous tubes are straight and distinguished from other regions of the breast by the name mammillary tubes.

One important result of the work of Ramsay *et al.* [85] is that the morphology of breast is not predictive of the internal anatomy of breast.

The morphology of the breast changes significantly due to changes in physiological conditions such as the menstrual cycle, pregnancy, and the onset of menopause. For example, during lactation, glandular tissue increases by two folds [88, 89, 90, 85]. Pathologies such as breast cancer also cause morphological

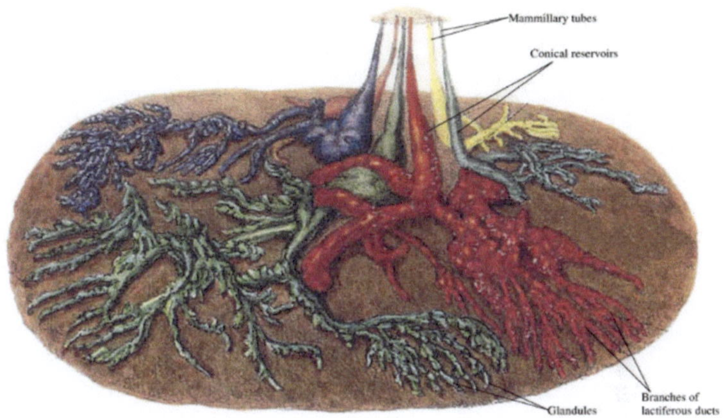

Fig. 3.2 A view of a preparation of six milk tubes injected from the nipple, showing the path from mammilary tubes to glandules (modified from [84]).

changes. As woman ages, connective tissues are replaced by fat and parts of the ductal networks are destroyed as ductal cells undergo atrophy [91].

Pathologies affecting the breast are typically categorized by the level of the ductal network in which they occur. Cancers originating from the lobules of the breast are called lobular carcinoma, while cancers origination from the ducts are known as ductal carcinoma. Rarely breast cancer can begin in the connective tissue that's made up of muscles, fat and blood vessels. That kind of cancer is called sarcoma.

Mammary ductal carcinoma is the most common type of breast cancer in women. It appears on the inner lining of the ductal networks. It takes two forms, ductal carcinoma in situ (DCIS), and invasive ductal carcinoma (IDC). DCIS is a non-invasive, possibly malignant, neoplasm that is confined to the milk ducts. DCIS develops into the more lethal form of breast cancer IDC, which invades the connective tissue network.

3.3 Elastomechanical Properties of the Breast

The mechanical properties of the breast are directly linked to the structure, composition and organization of tissues constituting the breast [92]. Quantifying the mechanical properties of breast tissues has been usually motivated by research aiming to identify ways for cancer detection and diagnosis [93, 94, 95, 96]. Therefore, a large number of these studies focused on evaluating the relative stiffness values of the different tissues and thus on estimating the Young's moduli associated with these tissues.

Commonly, the biomechanical properties of the breast are determined using uniaxial tension and compression tests on ex-vivo tissues samples [93, 94, 96]. Samani *et al.* [96] measured the mean Young's modulus E of 169 breast tissue samples by fitting displacement-force measurements of the samples to the equation:

$$E = \frac{\sigma}{\epsilon} = \frac{|\mathbf{f}|/A^0}{\Delta L/L^0} = \frac{|\mathbf{f}|L^0}{A^0 \Delta L} \tag{3.1}$$

where σ is the stress, ϵ is the strain, $|\mathbf{f}|$ is the force amplitude, A^0 the original cross-section area through which $\mathbf{f}$ is applied. L^0 is the original length of the object and ΔL is the displacement or change in object's length. Figures 3.3 shows a typical displacement-force measurement cycle for different tissue types.

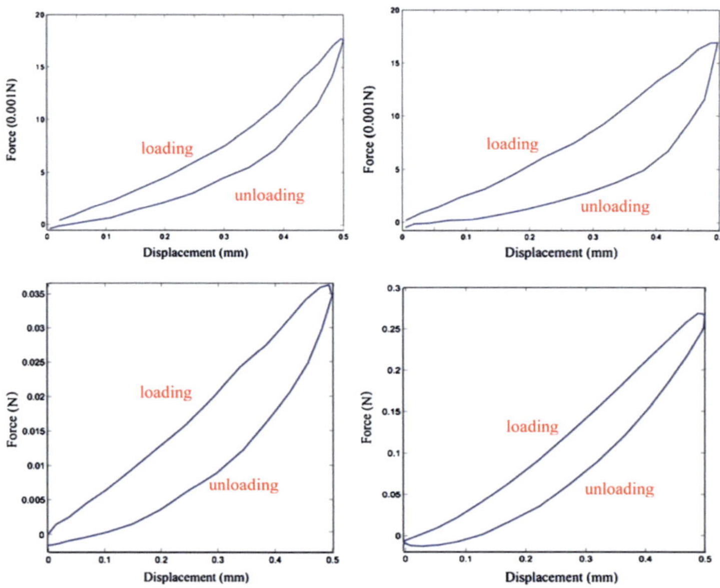

Fig. 3.3 Typical graphs of the force versus displacement cycles corresponding to normal fat tissue, normal glandular tissue, fibroadenoma and invasive ductal carcinoma (IDC) (Reproduced from [96]).

In general, ex-vivo measurements of breast mechanical properties [97, 93, 94, 96], and in-vivo measurements, based on magnetic resonance elastography [98, 95, 99], or on sonography [100] clearly indicate a wide variation in Young's Moduli values, not only among different types of tissues but also within each tissue type and was most evident in normal fat and fibroglandular tissues [101]. These variations can be traced back to changes to breast morphology with changing physiological conditions and age of a woman, and to inter-patient variability, and also to the methods used in measurements, leaving the door open for further investigation.

Krouskop *et al.* [93] showed that Young's moduli of breast tissues are highly dependent on the level of tissue pre-compression used in the measurement, confirming the non-linear elastic behavior often observed in biological tissues. Consequently, breast tissues exhibit a non-linear stress-strain relationship and can be considered to be elastic as there were negligible viscoelastic effects observed when compression tests were done at different strain rates. Hence Young's moduli are not sufficient to characterize the non-linear behavior of breast tissues, especially in cases of large-scale deformations.

Having an accurate mathematical model for the non-linear mechanical properties of the breast is essential for physically based modeling of breast deformation. The variation in mechanical properties of breast tissues across individuals poses a major challenge in developing anatomically realistic personalized breast models.

Gross pathological assessment and histological analysis confirm that tissue alterations such as breast cancer lead to macroscopic and microscopic structural reorganization. In general, abnormal and cancerous tissues are much stiffer than normal fat and fibroglandular tissues. Samani *et al.* found that the stiffness of fibroadenomas is approximately twice the stiffness of fibroglandular tissue, while infiltrating ductal carcinomas are thirteen times stiffer than normal fibroglandular tissue [96].

3.4 Modeling Breast Elastomechanics

Modeling breast elastomechanics is an active field of research. Around the world, many research teams are investigating the use of breast modeling in various clinical applications. Predicting mechanical deformation of breast tissues during biopsy procedures [102, 103], modeling breast compression [104, 105, 106, 107], the registration of MRI and X-ray mammograms [108, 42], and the prediction of breast surgery outcome [109] are among the renowned breast modeling applications.

Deformable breast models vary depending on many factors: The geometry and mesh used, the different constitutive laws, the deformable modeling techniques and the boundary conditions, and the application specific solution scenario, are among the most important factors.

For instance, methods for generating synthetic models of the breast have been developed [110, 42]. These models are based on general observations and gross anatomy of the breast. However, the large inter-patient variability of breast shapes, sizes and tissues distribution limit their use in personalized modeling for clinical applications. Detailed 3D images of patients' breasts comprising the internal structures can be generated by segmenting MR images where skin, fat and fibrous tissues can be easily identified. Cooper's ligaments and the breast tissue to muscle borders are typically difficult to identify and thus to segment from MR images. Therefore, most geometry models of breast used in modeling breast deformation neglect these structures.

As mentioned in Section 3.3, most stress-strain studies report mean values of Young's moduli E of different tissues that constitute the breast. For that reason, linear and piecewise-linear models based on E values are often used to describe the mechanical behavior of the different breast tissue types modeled as isotropic and homogeneous [111, 112, 113, 114]. Reported Young's moduli vary considerably across studies, as mentioned earlier. Furthermore they are not sufficient to reproduce the non-linear behavior of breast tissues, especially in cases of large-scale deformations. Many research groups use polynomial [109, 107] and exponential [115, 112, 114, 107] constitutive laws. The Mooney-Rivlin and the related Neo-Hookean constitutive laws have also been used in many models [116, 109, 117, 106] specially because they can achieve better fits for nonlinear functions.

Modeling the skin interaction with the internal breast tissue is not an easy task. The skin is often modeled as an additional layer of 3D elements that surround the internal tissue elements, or by coupling 2D membrane elements representing the skin to underlying 3D internal breast tissues elements [118, 107]. Little is known about the accuracy of these models, therefore more investigations must be conducted [116, 107, 119]

Constitutive laws assume zero stress at zero strain. Almost all geometrical breast models were obtained with the breast hanging under gravity or resting on a plate, which is not a stress-free reference state but rather a loaded configuration. Obtaining the stress-free configuration is important for the correct use of finite deformation elasticity theory [25] because the prediction of large deformations is a non-linear problem. The stress-free configuration is used as a reference state from which deformation predictions can be determined. Therefore, if the loaded configuration was used as the reference state, errors in deformation prediction will arise. By assuming that the breast density is similar to that of water, Rajagopal *et al.* estimated the unloaded state of the breast by imaging the breasts while immersed in water, and presented a method for determining the unloaded configuration of the breast from a set of deformed geometries. The resulting breast model was called the breast in neutral buoyancy. (see Fig. 3.4) [25, 101, 117, 120]. Rajagopal

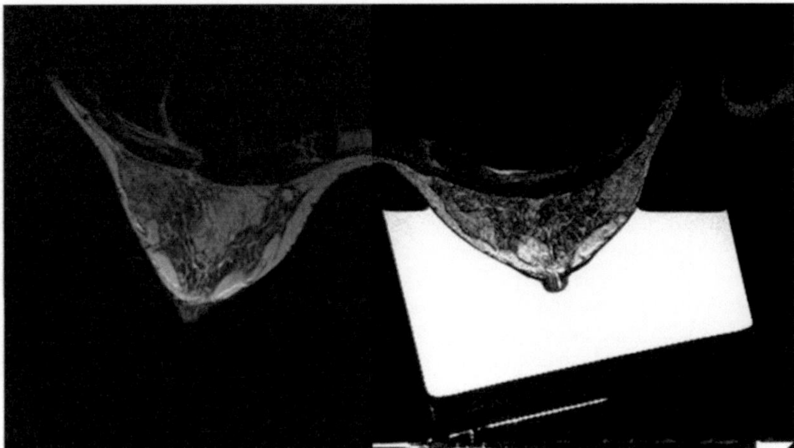

Fig. 3.4 Image of the breast in neutral buoyancy. The left image is of the left breast in the prone position. The right is the image of the right breast in neutral buoyancy. The white block represents the water bath.

showed also that it is possible to compute the reference state from a loaded state by redefining the knowns and unknowns in the problem definition for the Lagrangian formulation of the finite element implementation of the finite elasticity equations. The method was called the reverse method [101]. Rajagopal compared the breast in neutral buoyancy with the breast model generated using the reverse method, and concluded that the reverse method was more accurate in reproducing large and local deformations. A. del Palomar presented a different approach to obtain the stress-free configuration [109]. The model was submitted to gravity force pointing in the direction opposite to which during the breast image acquisition, in order to compensate the effect of gravity during image acquisition.

Most breast deformable models are based on the FEM method [112, 25, 106] where solutions are obtained using a static or steady state analysis. Other methods like the mass-spring method [110, 105, 42] and statistical deformation models [114] have also been used, with less success, due to the inherited limitations of these methods, especially the difficulty of enforcing volume preservation.

It is important to mention, that the breast is considered to be quasi-incompressible with Poisson ratio close to 0.5 [109]. In breast models, the breast is often rigidly fixed to the chest muscle, mainly because of lack of clarity about the muscle to the breast tissue boundary in MRI images.

Another difficulty is modeling the variety of complex loading conditions that the breast is submitted to, during the imaging procedures for example. The breast is subject to gravity-loading during MR imaging, however the breast comes in contact with the side of the MR coils what makes it likely that the breast is in that case subject to a more complex loading condition. Breathing makes the task even more difficult. Using the contact mechanics theory and a model for respiratory dynamics have to be used to increase the accuracy of modeling [119]. X-ray mammography clinical imaging conditions pose specific challenges. For instance, the breast is extended and placed on the x-ray mammography apparatus before compression [119]. The breast is then compressed between the plates. In almost all biomechanical models of mammography, the interaction between the plates and the skin of the breast is neglected, and the breast skin is assumed not to slide once they come in contact with the plates [102, 107, 113, 106]. During ultrasound imaging, the complex interaction between the sound probes and the breast tissue must be simulated dynamically [119].

To validate breast models, the modeling results are compared to mammograms or MRI datasets of volunteers [102, 103, 112, 113, 106, 109]. For example, displacements of predicted and marked internal feature positions are calculated [102, 103, 112, 113]. The displacement of skin markers was also used in the validation [109, 106]. Experiments with silicon gel phantoms to validate the modeling framework were also conducted in several studies [104, 102, 101].

From the above, it is easy to conclude that modeling breasts elastomechanics is not an easy task, and that it involves making various decisions about all the aspects mentioned in this section and other less important aspects not mentioned here.

3.5 Breast Elastomechanics under Gravity Loading

In this section the data and methods used for simulating the deformation of several breast phantoms with accurate dielectric properties from prone to supine positions are presented. The simulation scenario used for this task is then detailed and at the end the results of the simulation are presented.

A collection of anatomically realistic 3D numerical breast phantoms of varying shape, size, and radiographic density derived from T1-weighted MRIs of patients in prone position was developed by the group of C. S. Hagness [78] and made available to the scientific community through an online repository [121]. These phantoms capture the structural heterogeneity of normal breast tissue and incorporate the realistic dispersive dielectric properties from 0.5 to 20 GHz reported by Lazebnik [77, 122]. They are classified according to their radiographic density into four classes defined by the American College of Radiology [123]: (I) almost entirely fat, (II) scattered fibroglandular; (III) heterogeneously dense; and (IV) extremely dense. Figure 3.5 shows cross sections of breast phantoms each belonging to one of the four classes.

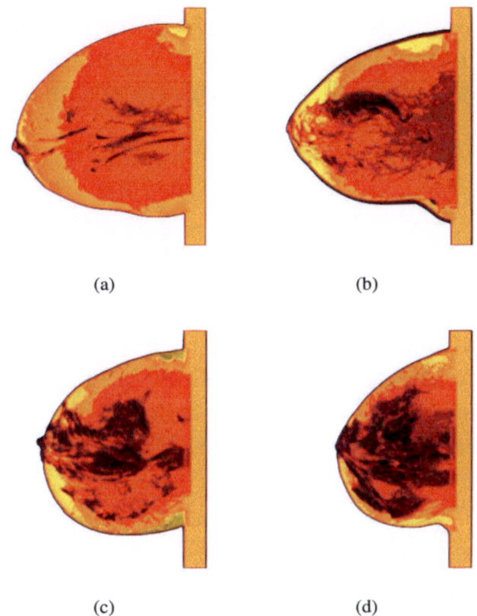

Fig. 3.5 Cross section of several breast phantoms belonging to the different breast classes: Breast ID 71904 mostly fatty breast (a), Breast ID 12204 scattered fibroglandular (b), Breast ID 80304 heterogeneously dense (c), Breast ID 12304 extremely dense (d).

The phantoms are provided in cubic voxel format with $0.5mm \times 0.5mm \times 0.5mm$ resolution. They include a $1.5mm$-thick skin layer, a $15mm$-thick subcutaneous fat layer at the base of the breast, and a $5mm$-thick muscle chest wall.

The normal breast tissue is categorized in the breast phantom into seven tissue types, ranging from the highest-water-content fibroconnective/glandular tissue with the highest dielectric properties to the lowest-water-content fatty tissue with the lowest dielectric properties, in addition to a transitional region with intermediate dielectric properties.

All these segmented breast datasets were obtained in prone position. Hence under gravity loading. Any geometry model created using one of these datasets will represent a breast deformed under the effect of gravity loading. To get the model in supine position, a stress-free reference model of the breast must be first calculated.

Several constitutive laws for the tissues present in the phantoms were investigated for this simulation task, including Mooney-Rivlin, Neo-Hookean [1] and the stress-strain relation described in the work of Azar *et al.* [102]. After evaluating a series of simulations, the Neo-Hookean constitutive law for a nearly incompressible isotropic material was chosen for all used material types. The used Neo-Hookean strain energy function can be derived from a modified Mooney-Rivlin strain energy function formulation for a nearly incompressible isotropic material. The modified Mooney-Rivlin strain energy function is given by

$$W = c_1(I_1 - 3) + c_2(I_2 - 3) \tag{3.2}$$

$$\widehat{W} = W + p_0 \ln I_3 + \frac{1}{2}\beta(\ln I_3)^2 \tag{3.3}$$

where I_1, I_2 and I_3 are the first, second and third principal invariants of the left Cauchy-Green deformation tensor, c_1 and c_2 are material related constants, β is a penalty parameter and must be large enough so that the compressibility error is negligibly small. The constant p_0 is chosen so that the components of the resulting Piola-Kirchhoff stress tensor $\mathbf{S}$ are all equal to zero in the initial configuration, i.e. $p_0 = -(c_1 + 2c_2)$. The Piola-Kirchhoff stress $\mathbf{S}$ can be written as

$$\mathbf{S} = 2\frac{\partial \widehat{W}}{\partial \mathbf{C}} = 2\frac{\partial W}{\partial \mathbf{C}} + 2(p_0 + \beta\ln I_3)\mathbf{C}^{-1} \tag{3.4}$$

where $\mathbf{C}$ is the left Cauchy-Green deformation tensor. The Neo-Hookean strain energy function for hyper-elastic nearly incompressible isotropic materials can be derived from the modified Mooney-Rivlin strain energy function by setting $c_2 = 0\,kPa$ in the mentioned equations. $\beta = c_1 \times 10^6 kPa$ was chosen for all simulations to ensure a relatively small change in tissues volume during deformation.

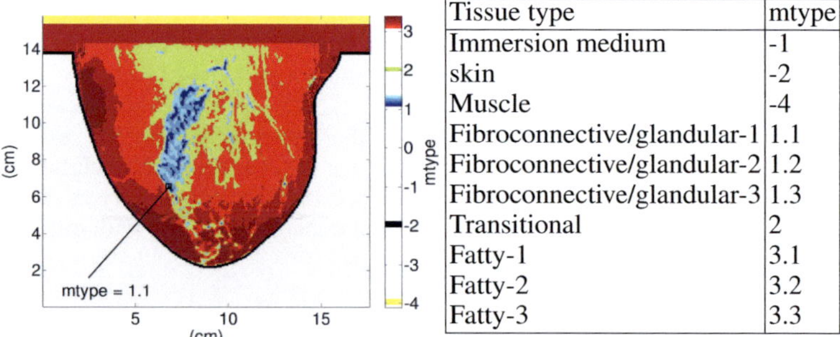

Tissue type	mtype
Immersion medium	-1
skin	-2
Muscle	-4
Fibroconnective/glandular-1	1.1
Fibroconnective/glandular-2	1.2
Fibroconnective/glandular-3	1.3
Transitional	2
Fatty-1	3.1
Fatty-2	3.2
Fatty-3	3.3

Fig. 3.6 Example of the spatial distribution of media numbers (provided in mtype.txt) in a sagittal slice from a 3D numerical breast phantom (Breast ID 12204) (left). and a table showing tissue types and corresponding media numbers (right).

Tetrahedral mesh topology based on a regular grid was used to the structure initialization of the modeling framework Adamss (see Section 2.3.1.3). Here all modeled tissue types are considered to be isotropic. The constitutive law in Eq. (3.3) was implemented and used with the method of the virtual hexahedron in the simulations (see Section 2.4.2.2). Linear friction of $2 \times 10^4 N/ms^{-1}$ was also introduced to the system. The Backward Euler method with adaptive time stepping and maximal time step of $h_{\mathrm{max}} = 4 \times 10^{-4} s$ was used for time integration of the equations of motion (see Section 2.5.5). The drop of the system's kinetic energy E_{k} below the threshold $E_{\mathrm{k,min}} = 0.001\ J$ was chosen to be the stop criterion for all conducted simulations of mechanical deformation.

Several boundary conditions setups were evaluated for the simulations. For instance, a strong fixation of the muscle layer at the base of the breast was first considered. This fixation restricted the deformation of the breast and resulted in unrealistic deformations. A restriction of the movement of this muscle layer in the direction normal to the chest wall was then introduced. This setting resulted in somewhat better deformations. However, due to the stiffness of the muscle layer, the breast formed an unrealistic nub. At the end, the muscle layer was omitted and a restriction of the movement of the fat layer in the direction normal to the chest wall was chosen as it gave the most realistic deformation among the evaluated set of boundary conditions. Skin was modeled as an extra 3D layer having the same elastic properties of fat tissue.

In this work a simulation scenario of two steps was used to simulate the deformation of the breast phantoms from prone to supine positions. At the end of the first step stress-free breast models are obtained. The stress-free models are used in the second step to obtain breast models in supine position. Both steps will be detailed in the following.

In the first step, stress-free models of the simulated breasts were obtained using the method introduced by A. del Palomar [109].

Basically, this method tries to reverse the effect of gravity loading during image acquisition by compensating gravity with force working in the opposite direction. To do that, the breast phantom generated in prone position is loaded with force equal to gravity but pointing towards the chest wall and the mechanical deformation is simulated using a set of constitutive laws parameters. At the end of the simulation a deformed breast model is generated. To make sure the deformed breast model is the stress-free model, the deformed breast mode is loaded with gravity pointing away from the chest-wall and a mechanical deformation is simulated using the same set of constitutive laws parameters. At the end the resulting model is compared with the original breast phantom. If the resulting model differs from the original, the procedure is repeated in an optimization loop, using different constitutive law parameters each time, until differences are minimal. Eventually, the stress-free model was generated by loading the original breast model with gravity pointing towards the chest and performing a mechanical deformation simulation using the found constitutive law parameters.

In this work, the optimization process shown in Figure 3.7 was used to obtain the suitable Neo-Hookean parameters and eventually the stress-free models.

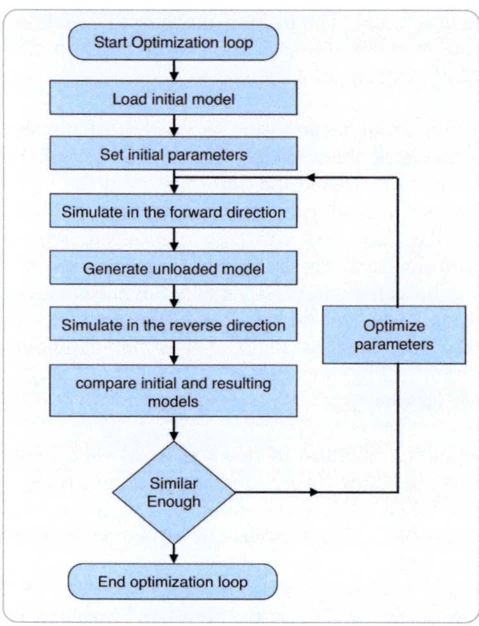

Fig. 3.7 Optimization loop for the Neo-Hookean parameters needed to obtain the stress-free models

To simplify the difficult task of finding Neo-Hookean parameters in the optimization loop, the same Neo-Hookean parameter was set to all three types of fibroconnective/glandular tissue present in the phantoms. The same applies for the different fat tissue types.

As shown in Figure 3.7, initial values for the Neo-Hookean parameter c_1 for Fibroconnective/glandular and fat tissues were first chosen (see Eq. 3.3). Then, the original model was simulated with gravity load pointing towards the chest-wall, until the stop criterion was reached.

As mention above, the tetrahedral mesh topology based on a regular grid was used to perform the modeling task using the modified mass spring system. Using this topology, each of the models consisted of about 80 millions tetrahedral elements. Even though the simulation framework used is of $O(n)$ complexity, simulating that amount of volume elements is time consuming and commands for a large amount of physical memory. Due to these limitations, and taking into account the quality of the results provided by the mechanical simulations, the datasets were first scaled down to $2mm$ grid resolution. Figures 3.8(a,b) show a snap shot of the original phantom in $0.5mm$ resolution and the downscaled model in $2mm$ resolution.

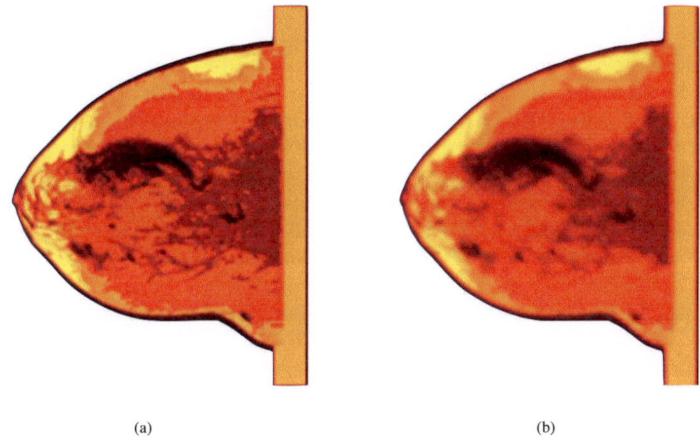

(a)　　　　　　　　　　　　　　　　　(b)

Fig. 3.8 2D cross section of breast ID 012204 in $0.5mm$ resolution (a), and the resulting model after scaling to $2mm$ resolution (b)

As a result of downscaling, the deformed models obtained after each simulation were of $2mm$ resolution. Re-meshing to a regular grid was used to turn the deformed models to stress-free deformed models.

To obtain a better re-meshing accuracy, the $2mm$ resolution deformed models were interpolated to $0.5mm$ resolution. Eight points interpolation [1] was used to interpolate the offsets of voxels' vertices of the deformed model from the $2mm$ to the $0.5mm$ resolution. Vertices in the $0.5mm$ resolution grid not belonging to any voxel of the $2mm$ grid were interpolated using the eight points linear extrapolation of the vertices of the nearest immediate neighboring voxels of the $2mm$ resolution grid.

Then, re-meshing according to a regular grid was used to obtain the stress-free deformed models starting from the interpolated $0.5mm$ resolution deformed models. The method used for re-meshing can be summarized in the following:
In each voxel of the original undeformed model, offsets associated with that voxel's vertices were used to determine the deformed version of that voxel. Each deformed voxel was then compared to a $0.5mm$ regular grid, that will eventually contain the stress-free model. In each voxel of that grid, a histogram of tissue types was set. Deformed voxels overlapping with voxels of the regular $0.5mm$ update the histograms of those voxels depending on the deformed voxel's underlying tissue type. In every voxel of the regular grid, the tissue type representing the maximal hits of that voxel's histogram was assigned to that voxel. By iterating over all regular grid voxels the stress-free models were generated.

Afterwards, the stress-free deformed model was used to start a reverse simulation using the same Neo-Hookean parameter values but with gravity loading pointing away from the chest-wall. The simulation runs until the stop criterion is reached. If the original model and the resulting models were not similar enough, the simulation was restarted with an updated c_1 value, as shown in the flowchart in Figure 3.7. And this procedure was repeated, until the difference between the original and the resulting models was minimal. The distance between the nipple and the chest-wall of the models was used as a similarity measure.

After several iterations in the optimization loop, the constitutive law parameters $c_1 = 0.2\,kPa$ and $c_1 = 0.8\,kPa$ were chosen for fat and fibroconnective/glandular tissue respectively. These values were used to generate deformed models of $2mm$ resolution. By interpolating and re-meshing, stress-free models of all simulated phantoms in $0.5mm$ resolution were obtained. The second step is straight-forward. Here, the stress-free models, resulting from the first step, were scaled down to $2mm$ grid resolution. Then, simulations of gravity loading in the direction towards the chest-wall were conducted using the parameters $c_1 = 0.08\,kPa$ and $c_1 = 0.32\,kPa$ for fat and fibroconnective/glandular respectively [101, 109, 119]. In these simulations, the same boundary conditions and the stop criteria used in the first step were used. By interpolating the offsets resulting from the simulation to $0.5mm$ resolution and re-meshing using a regular $0.5mm$ grid, the $0.5mm$ resolution models in supine position were eventually obtained.

Table 3.1 lists the conducted simulations and the corresponding calculated distances between the nipple and the chest wall in prone (d_{prone}) and in supine (d_{supine}) positions. The ratios $r\% = d_{\mathrm{supine}}/d_{\mathrm{prone}}$ are also inserted in the table. The calculated distances of the simulations correspond with measured distances reported in the literature. Additionally, the distances make the resulting breast models in supine position suitable for microwave imaging systems research.

Table 3.1 Breast phantoms simulations results

Breast ID	Description	d_{prone} (cm)	d_{supine} (cm)	$r\%$
071904	Mostly Fatty	13.18	6.35	46.02%
012804	Mostly Fatty	10.99	5.15	46.86%
012204	Scattered Fibroglandular	12.56	6.05	48.17%
080304	Heterogeneously Dense	10.54	6.02	57.12%
012304	Very Dense	8.08	4.74	58.66%

To obtain the dielectric parameters models in supine position, the offsets resulting from the first step (leading to the stress-free model) and the second step (leading to the model in supine position) were used in two consecutive re-meshing steps. In other words, the offsets resulting from the first step of the mechanical modeling were applied to the dielectric parameters models and re-meshing was performed and the dielectric parameters models in stress-free state were calculated. By applying the offsets resulting from the second step of the mechanical modeling simulations and re-meshing the dielectric parameters models in supine position were obtained.

Figures 3.9(a-c) show the original dielectric parameters of one of the breast phantoms taken in prone position, the stress-free model and the model in supine position resulting from the simulation. The figures show the flattening of the breast in supine position and the deformation of the internal tissue.

3.6 Discussion

In Section 3.5, a method to model the deformation of breasts from prone to supine positions under gravity loading using the modified mass-spring system was presented. Breast models with detailed dielectric properties for a wide range of frequencies of women in supine position were generated. The calculated breast models thicknesses ranged between $4.5 - 6.5cm$ (see Table 3.1), putting most of the volume of the breast to be within the range of currently available radar-based microwave imagine systems [81, 82] and making the models suitable for developing microwave imaging systems based on the supine position configuration.

Although, the original breast phantoms provided by Zastrow *et al.* [78] extracted from MRI images, captured the shape, size and structural heterogeneity of the normal breast tissue, an anatomically realistic chest-wall and muscle layer were not included. Instead an artificial $5mm$-thick cuboid-shaped layer of muscle tissue was added to the phantoms.

The muscle on top of the chest wall is the layer on which the breast deforms in supine position. It plays therefore an important role in the outcome of the deforma-

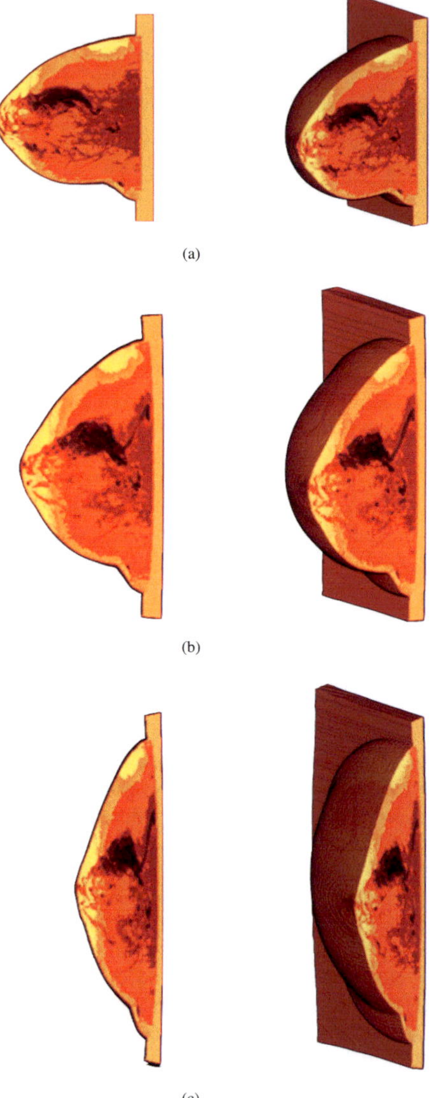

(a)

(b)

(c)

Fig. 3.9 Cross section of breast ID 012204 showing the heterogeneous dielectric properties distribution. The original phantom in prone position (a), the stress-free model (b) and the model in supine position (c).

tion. Little is known about the characteristics of the interface between the pectoral muscles and the internal breast tissues. Studies must be conducted to define the interface and to determine the best way to model the breast near the chest wall [116, 101]. Details about this interface could not only help improving the results of modeling breast deformation in supine position but also in other positions, like the supine-oblique position. Elevating the arms in supine position flatten further the breast and immobilizes it by tension from the pectoralis major muscle [83]. Including the contraction state of the muscle is crucial for reproducing a correct deformation.

The faithful representation of the heterogeneity in the breast is also challenging. When segmenting MR images, detailed 3D images of the internal structural organization of an individual's breasts, with a good contrast between skin, fat, fibrous tissues can be generated. Nonetheless, Cooper's ligaments and the breast tissue to muscle borders are typically difficult to identify and segment from MR images. Histological samples, such as those from Guinebretiere *et al.* [124] may help in developing models of anisotropy in the breast. But because of the lack of a clinical imaging protocol able to highlight these structures individual geometry models of breast used in breast deformation modeling neglect these structures. In this work for example, two tissue types, namely fat and fibroconnective/glandular, were modeled as isotropic Neo-Hookean materials and assigned different mechanical properties, neglecting Cooper's ligaments, the breast to muscle tissue borders and the skin, which was modeled as fat.

A tetrahedral mesh topology based on a regular grid was used in the simulations. This was essentially motivated by the fact that the original phantoms datasets are voxel datasets. To model smooth surfaces using this topology, a high resolution mesh is needed. However, to satisfy memory and computation time limitations, and at the same time minimally affect the mechanical behavior, the models were scaled down to $2mm$ resolution. At the end of the simulations a linear interpolation to a $0.5mm$ resolution and a re-meshing step were used to generate the resulting models in the original resolution. The errors generated from using a lower resolution, interpolation and re-meshing could be avoided using a better mesh topology such as the tetrahedral mesh topology based on an unstructured grid.

The mechanical properties of breast tissues differ between individuals and over time due to the variability in breast morphology, age, and physiological condition [125]. Such variability makes it difficult to use statistical data to model individual breast properties.

Williams *et al.* [126] conducted one of the few in-vivo compression experiments on the human breast to characterize the mechanical behavior of normal breast tissues. This method requires further extensions for more accurate characterization of mechanical properties. Rajagopal *et al.* [117] presented a method to acquire individual-specific Neo-Hookean material parameter values by using an optimizing technique to search for the parameter value that allows the biomechanical model to accurately reproduce a number of gravity-loaded configurations of the breast acquired using MRI. This technique requires further development to individually characterize the different breast tissue types.

Very few studies have been conducted to characterize the mechanical properties of the breast in-vivo under a wide range of strains and loading conditions. Future studies on characterization of mechanical properties of breast tissue should ensure that the nonlinear behavior of the tissues are captured and possibilities for in-vivo mechanical testing must be conducted ([101, 117]).

The Neo-Hookean constitutive law was used to model hyperelastic tissue types represented in the phantoms. The Mooney-Rivlin constitutive law given in Eq. (3.2) can achieve better fits for nonlinear models because of their greater number of free parameters. However, Tanner *et al.* concluded that using either constitutive laws does not influence the accuracy of the breast model [116].

Modeling the mechanical interaction of the skin with the internal breast tissue is not trivial. The simplest options involve modeling the skin as an additional layer of 3D elements that surround the internal tissue elements, or coupling 2D membrane elements representing the skin to underlying 3D elements representing the internal tissues of the breast [118, 107]. It follows from sensitivity analysis that little is known about the accuracy of these representations and that further investigation is needed [116, 107, 119]. In vivo measurement of the anisotropic properties of the patient's skin is also a challenging task. It requires multi-axially stretching of the skin. To perform these measurements, specialized devices that can be used clinically must be developed [119]. Therefore personalized skin modeling is currently off the table when modeling breast deformation.

Small and average size breasts flatten uniformly in supine position [83]. Larger size breasts tends to deform in the lateral direction in supine position forming folds. Modeling these deformations requires the implementation of collision detection and collision response methods in the mechanical modeling framework.

In Section 3.4, the importance of using stress-free breast models for modeling breast deformations under gravity loading was denoted. In this work, the method presented by A. del Palomar *et al.* [109] was used to create stress-free models of the breasts. This method requires a large number of parameter optimization steps, and thus a large number of mechanical deformation simulations which is computationally expensive. This computational cost is exposed to increase if the dimension of the parameters space increases by adding more tissue types or using more complicated constitutive laws. Furthermore, the similarity measure used with this method in this work is very simple. A better similarity measure must be used to ensure better parameters optimization. This will also contribute to an increment of computational cost as finding a local or global similarity maximum becomes more difficult. All these reasons make the use of the reverse method presented by Rajagopal [101] a much better choice, specially because it is mathematically well-founded and because it was validated in several experiments that showed its reliability, specially when more deformed states of the same breast are used to retrieve the stress-free model. Rajagopal called this the "reverse optimization method".

The validation of the resulting breast models in supine position was not possible since no information about the geometry of the simulated breasts in supine position was provided by Zastrow *et al.* [78]. To validate the method though, it is possible to do a validation study by recording the coordinates of surface markers arranged in a grid on the breast, in both prone and then in supine positions, and then calculate the distances between the markers' coordinates in supine position and those simulated. Additionally, by comparing segmented datasets of the breast in supine position with the simulation results, the method can be evaluated with respect to the internal features of the breast.

Despite all these issues concerning the modeling method presented in this work, the resulting breast models can be used to support the design of a Microwave Breast Scanner using numerical field calculation.

Chapter 4
Heart Modeling with Adamss

4.1 Motivation

Despite the significant medical advances over the last few decades, cardiovascular diseases (CVD)s remain the number one cause of death globally according to the World Health Organization (WHO) [64]. In 2004, around 17.1 million people died from CVDs representing 29% of all global deaths in that year, 7.2 million of which were due to coronary heart diseases and 5.7 million were due to stroke. These figures are projected to increase in the coming decades, keeping the CVDs the leading causes of death [64].

CVDs encompass a number of specific illnesses such as coronary heart disease, heart failure, arrhythmias, stroke, arterial and pulmonary hypertension, congenital heart disease, cardiomyopathies, valvular heart disease, etc.

Research focusing on the causes, diagnosis, treatment, and prevention of CVDs is moving ahead rapidly. For instance, more than $300 \in$ millions has been made available to support 39 EU-funded CVD research projects over the period 2002-2009 [127].

Computer modeling of the heart is valuable in understanding the functioning of the normal or diseased heart. The predictive potential of computer modeling of the heart is increasing rapidly. As new data on molecular and cellular mechanisms accumulates over time, the potential of heart modeling applications is growing [128], paving the way for better diagnostic strategies and new therapeutic approaches.

Many computer models of cardiac electrophysiology and mechanics have been developed and used in a range of applications like assessing the size and location of ischemic regions in heart of patients suffering from ischemia [129, 130], developing virtual surgery platforms for training or surgery planning [131, 132, 133, 134, 135, 136], assessing the effects of electric shocks or pacing [137, 138, 139, 140], and testing various hypotheses on heart physiology that cannot be tested in the real world due to physical or ethical reasons, such as whether the cardiac contraction remain homogeneous despite physiological asynchrony of depolarization [141].

A literature review published by Kerckhoffs *et al.* [142] presents a comprehensive overview of computer heart models, that witnessed a boom in the last two decades due to technological advances in computational power.

The generation of an accurate functional computer model of the heart with high computational efficiency and the ability to modify the model in order to reflect clinically observed phenomena, and to personalize it according to specific patient geometry, elastomechanical and electrophysiological characteristics, is the holy grail of cardiac computer modeling.

Towards that goal, the modified mass-spring system (Adamss) presented in Chapter 2 is used to model ventricular elastomechanics. Hereby, Adamss is used in combination with other methods to generate personalized anatomical models of the ventricles, to model cardiac electrophysiology represented by electrical excitation and excitation propagation, and to model the tension developed in cardiac ventricular tissue as a response to electric excitation.

Personalized modeling of the heart is a very challenging task. The anatomy of the heart shows large variations among individuals. It also shows changes within images of the same individual taken at different times, making the generation of accurate personalized anatomical models a difficult task. The electrophysiology and the elastomechanics of the heart are even more complicated. Some aspects of the heart physiology are yet to be explained. And in many cases there are no in-vivo methods for the measurement of essential parameters needed for the mathematical modeling of physiological processes. Therefore, a wise trade off between accuracy, efficiency and realizability must be made.

In the following sections, the anatomy and physiology of the heart are briefly introduced. A method for personalized modeling of the ventricles, is then detailed. The simulation setups and results are presented. At the end the methods and the simulation results are discussed.

4.2 Cardiac Gross Anatomy

The heart is a muscular organ located in the lower part of the left side of the chest (the thoracic cavity) and enclosed in a fibrous sac called the pericardium. In adults, it is roughly the size of its owner's clenched fist and weights between 230 to $350g$. The upper side of the heart is called the base, while the lower side is called the apex. Two thirds of the heart's mass lies to the left of the body's mid-line and its base-apex axis is tilted to the left [143]. Figure 4.1 depicts the anatomy of the heart.

The heart is part of the cardiovascular system that delivers nutrients including oxygen to cells, and removes metabolic waste products. The main function of the heart is to pump blood throughout the circulatory system by repeated, rhythmic contractions. It is divided longitudinally into two functional halves, left and right. Each half contains two chambers, an upper chamber called atrium, and a lower chamber called ventricle. These left and right halves are separated by the septum.

Blood is pumped from the right ventricle of the heart through the lungs and then to the left atrium. This is the pulmonary circulation. Blood is then passed to the left ventricle and pumped from there to the rest of the body which is the systemic circulation that ends with the right atrium. In both circuits, vessels carrying blood away from the heart are called arteries, and those carrying blood back toward the heart are called veins.

The walls in different parts of the heart have different thicknesses, due to different pressure against which the muscle has to act. The walls of atria are less than $2mm$ thick, while the wall of the right and left ventricles has approximately $10mm$ and $20mm$ thickness respectively. These values correspond with pressure differences of around $20mmHg$ in the atria, around $30mmHg$ in the right ventricle and around 80 to $120mmHg$ in the left ventricle [145].

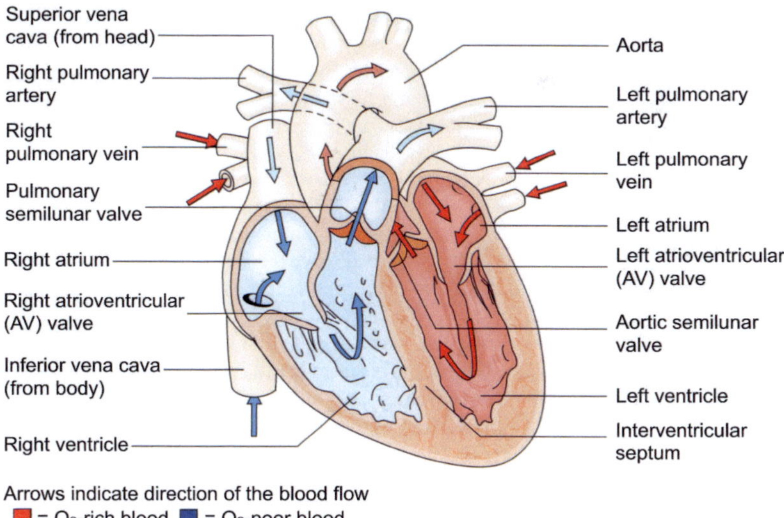

Fig. 4.1 The anatomy of the heart (reproduced from [144]).

The overall arrangement of the circulatory system is illustrated schematically in Figure 4.2.

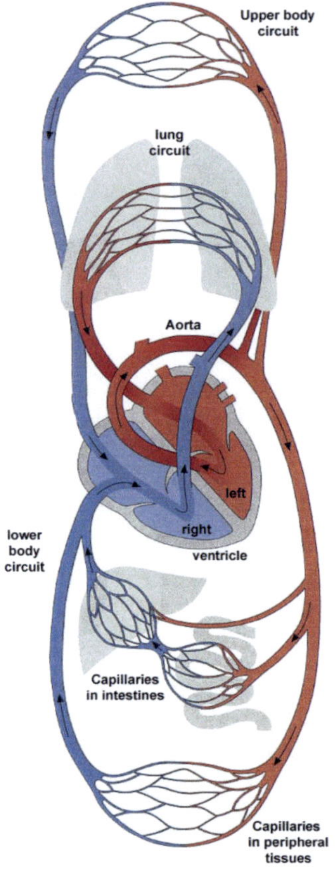

Fig. 4.2 Schematic drawing of the circulatory system (reproduced from [146])

The right atrium is attached to the vena cava superior and inferior, while the left atrium is attached to the venae pulmonalis. The right ventricle transports oxygen-poor blood through the arteria pulmonalis to the lungs, whereas the left ventricle is attached to the aorta through which oxygen-rich blood is transported to the rest of the body.

The coronary arteries, that start at the root of the aorta, supply the atrial and ventricular walls with blood.

Four dense connective tissue rings surround the valves of the heart, fuse with one another, and merge with the interventricular septum at the valves plane. This connective tissue is called the fibrous skeleton of the heart. In addition to forming a structural foundation for the heart valves, the fibrous skeleton prevents overstretching of the valves as blood passes through them, and serves as a point of insertion for bundles of cardiac muscle fibers and acts as an electrical insulator between the atria and ventricles [147].

The atrioventricular (AV) valves are located between the atrium and ventricle in each half of the heart. Each permits blood to flow from atrium to vectricle but not in the opposite directions. The right AV valve is called the tricuspid valve, and the left is called the mitral valve. The opening and closing of the AV valves is a passive process resulting from pressure difference across the valves. When blood pressure in an atrium is greater than that in the ventricle, the valve is pushed open and blood flow proceeds from atrium to ventricle. In contrast, when a contracting ventricle achieves an internal pressure greater than that in atrium, the AV valve is forced to close. Figure 4.3 shows a schematic representation of the valves plane.

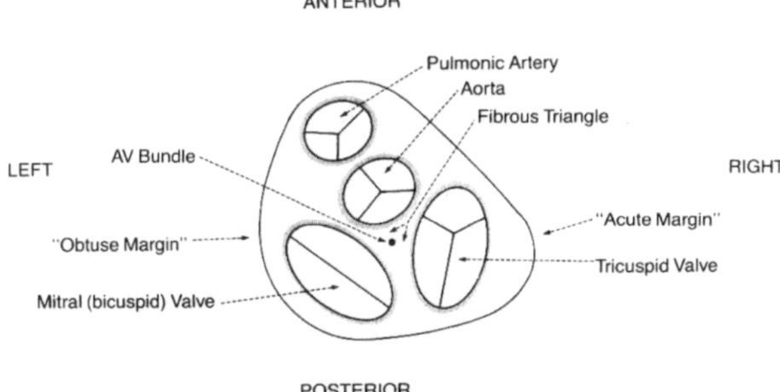

Fig. 4.3 Schematic representation of the valves plane showing the triscupid and the mitral valves connecting atria to ventricles and the aorta and pulmonary arteries openings (reproduced from [148]).

To prevent the AV valves from being pushed up into the atria, the valves are fastened by fibrous strands to muscular projections of the ventricular walls called papillary muscles. The papillary muscles do not open or close the valves. They act only to limit the valves movements and prevent them from being everted.

The opening of the right ventricle into the pulmonary trunk and of the left ventricle into the aorta also contain valves, called the pulmonary and aortic valves. These permit blood flow into the arteries during ventricular contraction and prevent blood from moving in the opposite direction, to the ventricles, during ventricular relaxation. As the AV valves, they also act in a purely passive manner.

There are no valves at the entrance of the superior and inferior venae cavae into the right atrium and of the pulmonary veins into the left atrium. However, atrial contraction pumps very little blood back into the veins because atrial contraction compresses the veins at their sites of entry into the atria, what greatly increases the resistance to backflow [149].

The walls of the heart consist of three layers. From the innermost to the outermost of the four heart chambers, these layers are the endocardium, the myocardium and the epicardium.

The endocardium is a thin layer of endothelium overlying a thin layer of connective tissue. And, it is continuous with the endothelial lining of the large blood vessels attached to the heart. It provides a smooth lining for heart chambers and covers the valves of the heart, hence minimizing the friction at the surfaces as blood passes through the heart.

The myocardium is composed of cardiac muscle tissue and makes up to 95% of the heart walls. It is responsible for the pumping action of the heart. The myocardium can be classified to subepicardium, midwall or subendocardial layers. The individual cells, or muscle fibers, that form the myocardium are called cardiac myocytes. The cardiac muscle is organized in a pattern of struts and weaves, criss-crossing bundles of muscular beams, that generate the strong pumping actions of the heart.

In comparison with skeletal muscle fibers, cardiac muscle fibers are shorter in length and less circular in transvers section. They also exhibit branching, giving individual cardiac muscle fibers a stair-step appearance as shown in Figure 4.4(a).

Cardiac muscle fibers are connected to neighboring fibers at the endings by intercalated discs containing a large density called of gap junctions. They allow actions potentials to propagate form one fiber to its neighbours.

In ventricles, three to ten cardiac myocytes are grouped together and surrounded by a network of connective tissue. These groups form multiple sheets that are loosely coupled by sparse and long collagen fibers building a mesh. The mesh, that surrounds and prevades the cardiac myocytes, makes about $2 - 5\%$ of the weight of the heart. Furthermore, fibers of elastin, are spanned in the different layers of the myocardium.

The cardiac connective tissue consists of collagen, elastin, glyosaminoglycans and glycoproteins [150, 151]. It can be divided into endomysium, perimysium and epimysium [2].

Endomysium surrounds the myocytes and binds the fibers of one sheet tightly together. Perimysial collagen weakly connects sheets together (Fig. 4.4(b)), giving the sheets the possibility to slide over each other with relative ease [152]. Finally the epimysium is a layer of collagen and elastin fibers that covers the subendo- and subepicardial myocardium [150, 151, 153, 23, 2].

Both the collagen and the elastin contribute to the viscoelastic behaviour of the myocardium and play an important role in the mechanical function of the ventricles as they define the passive mechanical properties of the tissue [154, 150, 152].

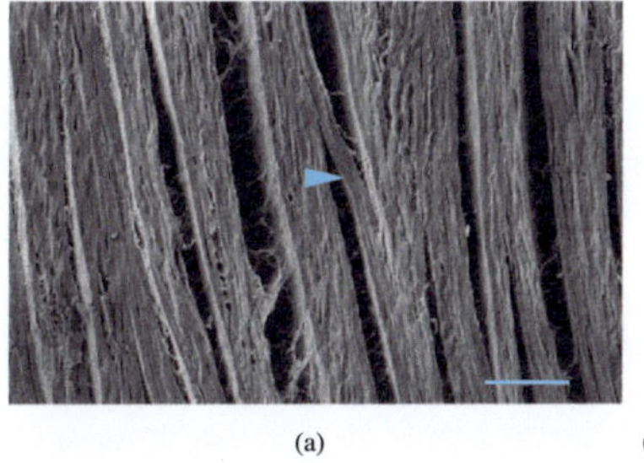

(a)   (b)

Fig. 4.4 Micrograph of myocytes branching (arrow) collagene fibers between adjacent sheets, scale bar, $100\mu m$ (a). Perimysial connective tissue weaves surrounding myocardial sheets, the scale bar at lower right corner represents $25\mu m$ (b). (figures edited from [152]).

The muscle fibers in the ventricles are oriented and laminated in an organized manner. Measurements of human left ventricular tissue made by Streeter [155] showed a continuous transmural rotation of the main helix angle, which is the angle of the fibers parallel to the local endocardial surface. If the orientation from apex to base was chosen to be $\pm 90°$ and the angle $0°$ the equatorial direction, then, according to the measurement of Streeter, the helix angle of the human left ventricle is $55°$ in the subendocardial border, and $-75°$ in the subepicardial layer as can be seen in Figure 4.5. Between the subendo- and the subepicardial layers the helix angle changes linearly with the value $0°$ near the middle of the left ventricular wall.

A linear change of the helix angle was also reported in the right ventricle.

At the apex and the middle of the septum, the fiber orientation is more complex. Papillary muscles have a longitudinal fiber orientation along their main axis.

Fig. 4.5 A reconstruction of a stack volume of the ventricular wall of a rat. The rotation of the fiber orientation from subepi- to subendocardium is easy to identify. (reproduced from [156]).

Using MR diffusion tensor imaging (MR DTI) it is possible to study fibrous structures like the brain, and muscles including the heart [157, 158]. Zhukov *et al.* [157] extracted the fiber orientation of an extra-corporal canine heart in contracted state. Figure 4.6 depicts the fiber orientation in both ventricles of a canine heart. Since

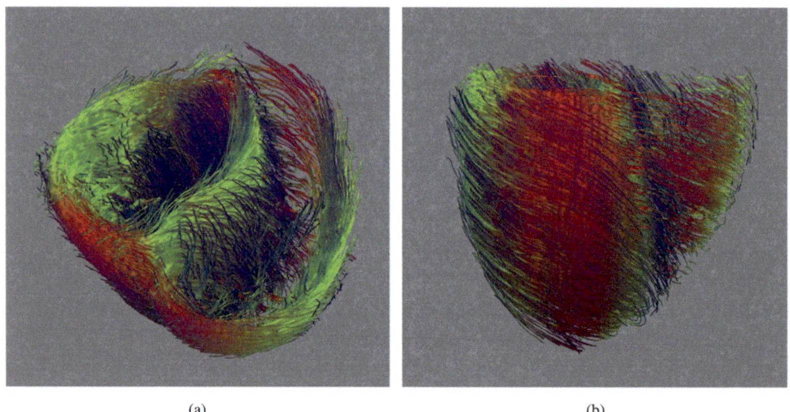

Fig. 4.6 Fiber orientation of a canine heart reconstructed using MR DTI imaging technique. The reconstruction shows the spiraling fibers orientation (modified from [157]).

the geometry of the atria is more complex than that of the ventricles, the fiber orientation of the atrial myocytes is complex. Ho *et al.* reviewed the general myocardium architecture of the atrial walls revealed by the arrangement of myofibers. The reader is advised to the work of Ho *et al.* [159] for more information.

The epicardium is composed of two tissue layers. The outermost is called the visceral layer of the serous pericardium, which in fact makes part of the pericardium. Beneath it, the second layer is found, which is a variable layer of delicate connective tissue or adipose tissue. The adipose tissue predominates and becomes thickest over the ventricular surfaces, where it houses the major coronary and cardiac vessels of the heart.

The heart lies within the pericardium, a tough fibrous sac attached to the diaphram that surrounds and protects it. The narrow space between the pericardium and the heart is filled with a watery fluid that serves as a lubricant as the heart moves within the sac.

The apex of the heart is attached to the pericardium with a realively strong bound. When the ventricles contract the heart atria move towards the apex and expand. The superficial fibrous pericardium is a tough, inelastic, dense irregular connective tissue that resembles a sac that rests on and attaches to the diaphram. The open end of the bag is fused to the connective tissues of the blood vessels entering and leaving the heart. The fibrous pericardium prevents overstretching of the heart, provides protection and anchors the heart in the mediastinum.

4.3 Cardiac Myocytes

Cardiac myocytes have commonly a cylindrical shape. Their length ranges between 50 and $120\mu m$, whereas their diameter ranges between 5 and $25\mu m$ [160]. Cardiac myocytes are enclosed by a cell membrane, called the sarcolemma, that separates the intra- and the extracellular spaces. The intracellular space of the myocyte contains the nucleus, mitochondria, myofibrils, the sarcoplasmic reticulum and the cytoskeleton that maintains the mechanical integrity of the cell [161].

The remaining volume of the intracellular space is filled with a liquid solution containing lipids, carbon hydrates, salts and proteins. The main components of the cardiac myocyte are illustrated in Figure 4.7.

The sarcolemma includes various pore forming proteins that take a cylindric shape with diameters of about $1nm$ [163]. These proteins are ionic channels, exchangers and pumps, specified by the ion type that can pass through like the sodium channel or the calcium channel. These proteins define the sarcolemmal permeability to a specific ion type.

Thousands of tiny invaginations of the sarcolemma, called transverse tubules, tunnel in from the surface toward the center of each muscle fiber. Transverse tubules, or in short form T tubules, are open to the outside of the fiber and thus

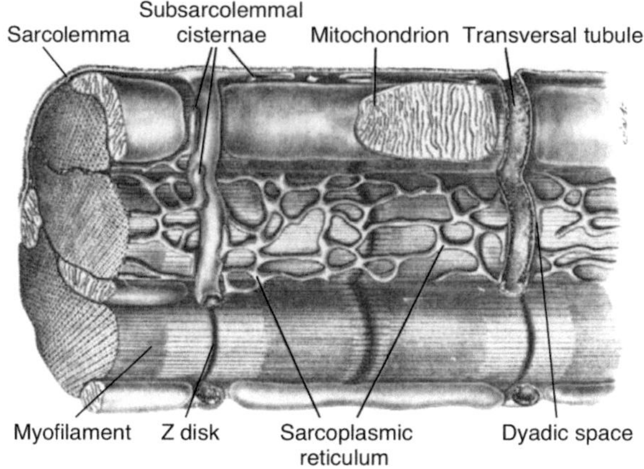

Fig. 4.7 A schematic description of a cardiac myocyte (adapted from [162]).

are filled with interstitial fluid. Action potentials propagate along the sarcolemma and through the T tubules, quickly spreading throughout the muscle fiber. This arrangement ensures that all the superficial and deep parts of the fiber become excited by an action potential almost simultaneously [147].

Mitochondria are cellular organelles with lengths ranging between 0.3 and $1.7\mu m$, and diameters ranging between 0.2 and $1\mu m$. They supply the cell with energy by metabolism, a function performed by its internal components via oxygenation, generating adenosine triphosphate (ATP). Because of their function as cellular energy provider, mitochondria are located near the energy recipients such as the myofilaments and the ion pumps that consume ATP. They occupy from 16% to 20% of the volume of the atria and from 25% to 36% that of the ventricles [164].

Myofibrils are approximately cylindrical bundles of an arrangement of numerous thick and thin filaments. 41% to 53% of the volume of atria, and 45% to 54% of the volume of ventricular cardiac myocytes is filled with myofibrils.

Thin and thick filaments in each myofibril are arranged in a repeating pattern along the length of the myofibril. One unit of this repeating pattern is known as a sarcomere. The sarcomere is the fundamental contractile unit within the cardiac muscle. Each sarcomere is about $2\mu m$ in length. When contracting, the length of a sarcomere can reach 60% of the sarcomere resting length.

Each myofibril is made up of millions of sarcomeres which are placed in a back to back arrangement. Thick filaments are composed almost entirely of the contractile protein myosin. Thin filaments contain the contractile protein actin as well as two other proteins, troponin and tropomyosin, that play an important role in regulating contraction.

Thick filaments are located in the middle of each sarcomere. Two sets of thin filaments are anchored to a network of interconnecting proteins known as the Z disc, whereas the other ends overlap portions of the thick filament. Two successive Z discs define the limits of one sarcomere.

The parallel arrangement of thick filaments produces a wide dark band known as A-band, when subject to polarized light. Between A-bands, regions, consisting of thin filaments appear in a light color. These are called I-bands. Each I-band is divided by a characteristic Z disc. Figure 4.8 depicts these structures.

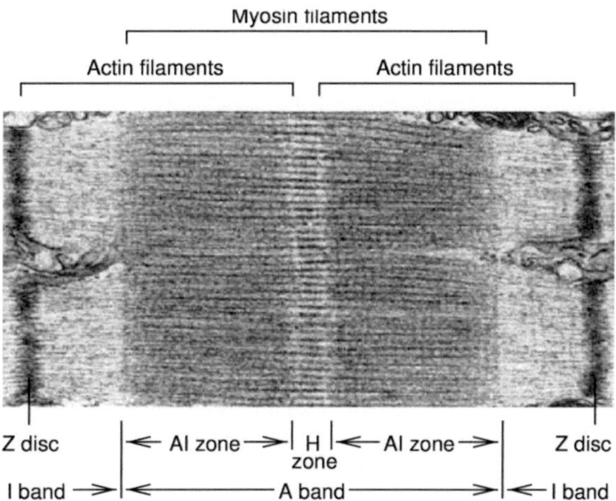

Fig. 4.8 Electron microgrphy of a sarcomere of a cardiac myocyte, showing the A and I bands and Z discs (adapted from [165]).

The sarcoplasmic reticulum (SR) is a membranous structure that enfolds the sarcomeres [166]. The SR membrane contains calcium, potassium, chlorine and hydrogen ionic channels as well as calcium pump proteins that regulate the myoplasmic concentration of calcium. SR serves as a high capacitance reservoir for calcium ions, which are important for the contraction process. They can be seen as an important modulator in excitation contraction coupling [167, 168].

4.4 Cardiac Electrophysiology

Since cardiac electrophysiology is tightly coupled with the mechanical contraction and thus the pump function of the heart, it will be briefly introduced in this section.

Generally speaking, cardiac myocytes are the origin of the electrical activity of the heart. Several types of myocytes exist in the myocardium. Some are specialized in the initiation of electrical activity, while other are responsible for the fast conduction of electrical excitation.

The majority of myocytes are responsible for generating the mechanical tension and thus the contraction of the heart. These myocytes generate tension by the electro-mechanical coupling mechanism. Furthermore, they conduct the electrical excitation to neighboring cells, primarily by transport of intercellular ions via the gap junctions.

4.4.1 Cell Membrane

Every cell is enclosed by a thin membrane, which serves as a barrier between the intra- and extracellular spaces. The cell membrane, shown in Figure 4.9, consists of phospholipid bilayer containing pores formed by proteins. The membrane, allows for the passive diffusion of hydrophobic molecules. However, ions like sodium (Na^+), potassium (K^+) or calcium (Ca^{2+}), whether outside or inside the cell, can cross the membrane only through ion-specific transport proteins. The difference in ion concentrations between intra- and extracellular spaces induces an electrochemical potential difference across the membrane [163].
The cell membrane plays a major role in the resting and active electric properties of excitable cells, like neurons and myocytes. The membrane has different permeabilities for different ions due to the different transport proteins and their states. The selective permeability alternations, and the regulation of ion currents generate the basic bioelectric phenomena [169].

Transport proteins can be divided according to their transport mechanisms to ion pumps, ion channels and ion exchangers.

4.4.1.1 Ion Channels

Ion channels are transmembrane proteins containing a central pore or tunnel. This pore allows ions to pass into and out of the cell down the concentration gradient. They are distinguished based upon their ion selectivity and gating mechanism.

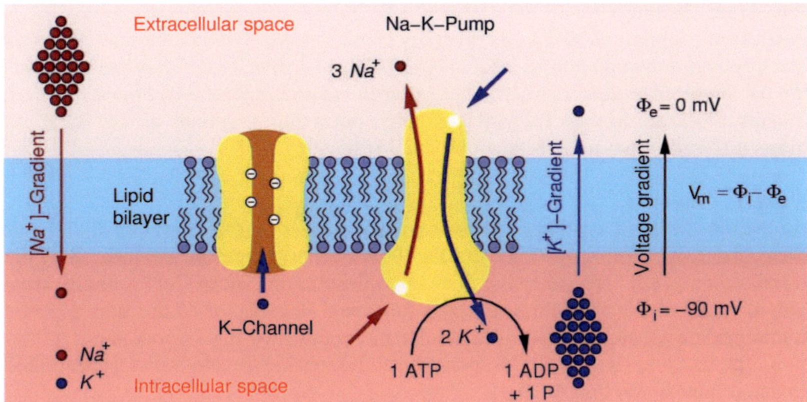

Fig. 4.9 Illustration of a cell membrane. Different concentrations of ions in intra- and extracellular spaces generate an electrochemical potential difference across the membrane. Ion channels and pumps control the transport of ions in and out of the cell (modified from [163]).

Ion channels can be in a conductive, a non-conductive, or an inactive state. Gating is the process of switching from one state to the other. Selectivity refers to the ability of the channels to allow the flow of only one specific ion type. With gating and selective permeability, ion channels regulate the ionic distribution in the intracellular space.

Voltage-gated channels change their protein configuration depending on the transmembrane voltage, switching from one state to the other [170, 171]. Other type of channels have binding sites for neurotransmitters. Gating in these channels is controlled by the neurotransmitters whether its origins were extracellular like adrenaline or intracellular ligands like ATP [172, 173]. Gating can also be triggered by mechanical stretch of the membrane [174] or ionic concentrations.

The most important channels for the cardiac activity are the voltage-gated potassium, sodium, and calcium channels discussed below.

Sodium Channels
Voltage-gated sodium channels are voltage-dependent [175] and are responsible for the fast depolarization of the cell membrane [170, 176]. If the transmembrane voltage exceeds a threshold the probability of transition from a non-conductive state to the conductive state increases strongly [177, 170]. Sodium channels open very fast and sodium flows into the cell. Only $0.1ms$ later they close and switch to the inactive state.

Potassium Channels
Potassium channels show a large diversity is their molecular arrangement and their electrophysiology and gating mechanisms [178, 179, 180, 181]. They are responsible for the repolarization of myocytes, after the action potential had been induced. Twenty types are known for voltages-gated potassium channels alone. Each exhibits different time- and voltage-activated behavior [171].

Calcium Channels
At least 4 different types of calcium channels exist in the heart [182]. Two of them, the L-type (long-lasting) and the T-type (tiny) channels are located in the sarcolemma [183]. The other two, the sarcoplasmic reticulum Ca^{2+} release channel, also called the ryanodine receptor, and the IP_3 receptor of the sarcoplasmic reticulum are located on the intracellular membranes of the sarcoplasmic reticulum. The calcium dynamics is important for the tension development as described in detail later in Section 4.7.4.

4.4.1.2 Ion Pumps

Ion pumps transport specific ions against their concentration gradient using energy in form of ATP. Pumps produce and maintain an electro-chemical gradient which is the driving force for the fast currents through ion channels [184].

Sodium Potassium Pump
The Na^+-K^+ pump transports two K^+ ions into the cell and three Na^+ out of the cell in one transport procedure consuming the energy of one ATP molecule. It is mainly responsible for maintaining the resting voltage and thus the excitability of the myocyte. The transport is found to be dependent on the concentration of intra- and extracellular sodium and potassium, the transmembrane voltage as well as on temperature [175].

Calcium Pump
The calcium pump transports Ca^{2+} from the cytoplasm to extracellular space. The process depends on the intracellular calcium concentration [160].

Sarcoplasmic Reticulum Calcium Pump
The sarcoplasmic reticulum calcium pump transports Ca^{2+} from the cytoplasm into the sarcoplasmic reticulum that serves as a calcium reservoir.[166].

4.4.1.3 Ion Exchangers

Ion exchangers exploit the concentration difference of one solute between the intra- and extracellular spaces to transport a second solute. This phenomenon is called co-transport when diffusion energy of the first solute is used to pull other

substances through the cell membrane. It is called counter-transport when both so-
lutes move in opposite directions through the cell membrane [160, 144].

Sodium-Calcium Exchanger The sodium-calcium exchanger plays an important
role in the cardiac electrophysiology. It regulates the Ca^{2+}-distribution in the intra-
and extracellular spaces, which is important for the contraction of the heart.

The sodium-calcium exchanger, transport calcium ions that passed into the cell
via the sarcolemma calcium channels and leak currents, out of the cell by exploit-
ing the high concentration of sodium ions in the extracellular space. The exchanger
transports one calcium ion from the cytoplasm for three extracellular sodium ions
that pass to the cell with the diffusion energy. The Na^+-Ca^{2+} exchanger can also
work reversely if the cell is depolarized sufficiently. In that case sodium is transported out of the cell and calcium is transported into the cell [160, 185].

4.4.1.4 Gap Junctions

At the intercalated disks level, mainly near the ends of cardiac myocytes, specialized proteins of adjacent cells form, in a bundled manner, a low electrical resistance pore called gap junction [186]. Gap junctions allow the transport of small
water soluble particles e.g. ions, without ever entering the extracellular fluid.

A notable distinction can be made between the longitudinal and transversal gap
junctions. Longitudinal gap junction are oriented, approximately, in the same direction as the first principal axes of adjacent myocytes. While transversal gap junctions can be oriented in the plane transversal to that axis.

The density and distribution of gap junctions differ depending on the type of
the tissue [187]. An increase in the gap junctions density in a region of the myocardium will result in an increased excitation velocity in that region.

4.4.2 Transmembrane Voltage

The different ion permeabilities and active ions transport give rise to differences in
concentrations of ions in intra- and extracellular space and thus to different electrical potentials across the membrane [160]. The resulting potential difference across
the cell membrane is called transmembrane voltage.

The voltage of the equilibrium state is called resting potential. It ranges from
$-30mV$ to $-100mV$ depending on the cell types. By considering only Na^+, K^+
and Cl^- ions, the resting potential can be described by the Goldman-Hodgkin-
Katz equation [169]:

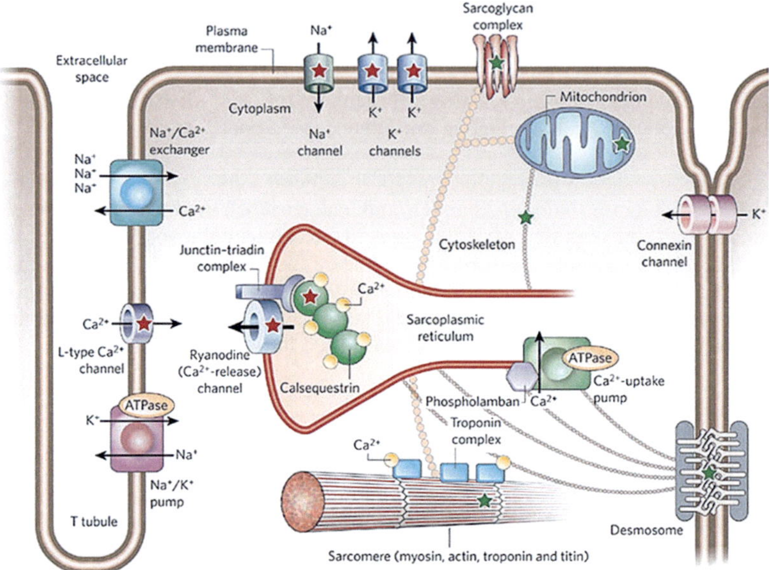

Fig. 4.10 Illustration of a ventricular cardiac myocyte. The main components of the cell and the different kinds of ion channels, exchangers and pumps are depicted (reproduced from [188]).

$$U = -\frac{RT}{F} \ln\left(\frac{P_K[K^+]_i + P_{NA}[Na^+]_i + P_{CL}[Cl^-]_o}{P_K[K^+]_o + P_{NA}[Na^+]_o + P_{CL}[Cl^-]_i}\right) \qquad (4.1)$$

where R is the gas constant, T the temperature in Kelvin and F the Faraday constant. P_X is the permeability of ion X and $[X]_{i/e}$ is its intra-, and extracellular concentration, respectively. At body temperature of $37°$ the resting potential is given with

$$U = -61 \ln\left(\frac{P_K[K^+]_i + P_{NA}[Na^+]_i + P_{CL}[Cl^-]_o}{P_K[K^+]_o + P_{NA}[Na^+]_o + P_{CL}[Cl^-]_i}\right) mV \qquad (4.2)$$

The permeability P_X can be derived from Fick's diffusion law and is given as follows:

$$P_X = \frac{D_X \beta_X}{h} \qquad (4.3)$$

where h is the thickness of the membrane, D_X the diffusion coefficient of the ion X and β_k the water-membrane partition coefficient.

The Goldman-Hodgkin-Katz equation does not consider the permeability of Ca^{2+} ions and the charge transport of the sarcolemmal pumps and exchangers.

4.4.3 Action Potential

In excitable cells, ion permeabilities of the cell membrane can change causing the development of an action potential. An action potential will not occur before the initial rise in membrane potential exceeds a certain threshold to create a vicious positive feed-back loop that opens the sodium voltage-gated channels in an avalanche-like manner [160].

Usually, an action potential starts from the resting stage, where the membrane is said to be polarized and has the resting negative membrane voltage. As an external stimulus or an impulse conducted through the gap junctions excites the cell. A sudden increase in the membrane permeability to sodium ions allows tremendous numbers of positivly charged sodium ions to flow into the cell and the potential rises rapidly in the positive direction. This is called the depolarization of the membrane. After approximately $0.1ms$, sodium channels switch to the inactive state and stay closed for a certain time. The calcium and the early potassium channels begin to open but much slower than the sodium channels. Then, rapid diffusion of potassium ions to the exterior re-establishes the negative membrane voltage, and the membrane is said to repolarize [160]. This change in the membrane voltage is termed action potential (AP). Figure 4.11 shows a typical action potential of a cardiac myocyte and the related ion permeabilities.

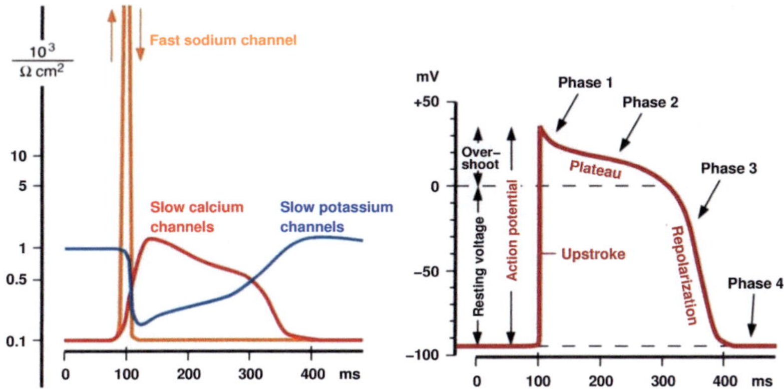

Fig. 4.11 The selective membrane ion permeabilities during an excitation of the cell (left) and an action potential labeled with related terms (reproduced from [184]).

After the sodium channels have switched to the inactivated state, the cell can not be excited for a certain duration. This period is called the absolute refractory period. It is followed by the relative refractory period in which the cell is not back yet to the resting stage, but sodium channels can partly be excited again. An AP triggered in this period, has a shorter duration and a lower amplitude as demonstrated in Figure 4.12.

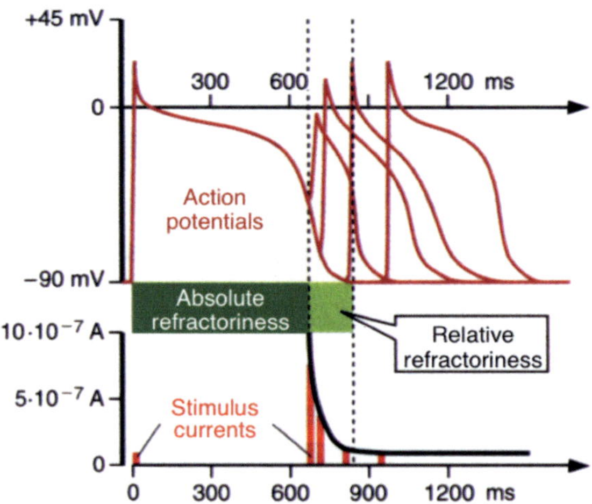

Fig. 4.12 Refractory period of a cell. During the absolute refractory phase a new action potential cannot be initiated. During the relative refractory period an action potential with shorter magnitude and duration can be stimulated (modified from [189]).

4.4.4 Electrical Conduction System

The electrical excitation propagates throughout the myocardium in a way such that it results in an optimized pumping function of the heart.

Cardiac myocytes are electrically coupled mainly by gap junctions that behave like low resistance ohmic resistors and also by the extracellular space they share. The excitation of one cell results in a voltage gradient between adjacent cells. This causes a current through the gap junctions as well as through the extracellular space which activates further cells.

The excitation conduction is mainly influenced by the current that flows through the gap junctions. The density and distribution of gap junctions as well as the shape of the myocytes depend on the structure of tissue and the fibers orientation. Therefore the myocardium shows a macroscopic anisotropic conductivity [190, 191].

The cardiac system of excitation initiation and conduction consists of specialized myocytes on the levels of cellular anatomy, intercellular coupling and the cellular electrophysiology.

The primary pacemaker of the excitation initiation system is the the sino-atrial node (SN), which is a small crescent-shaped tissue located in the wall of the right atrium [143]. The SN contains the fastest autorhythymic cells [192]. Once an AP is generated in the SN, it propagates throughout both atria via the atrial cardiac myocytes and specialized conduction pathways, finally reaching the atrioventricular node (AVN). The AVN is the single electrical contact between atria and ventricles. Cells of the AVN conduct the excitation with a delay enabling the atria to become completely depolarized and to contract before the ventricles.

After that, the AP goes down the septum via the His bundle. The His bundle is isolated by fibrous tissue and is the only physiological electrical pathway between the atria and the ventricles. It merges into the left and right Tawara bundle branches. The AP propagates further through the Tawara bundle branches and the subendocardial network reaching throughout the ventricles (Purkinje fibers). Both ventricles get activated from endocardium to epicardium and from apex to base, and they contract thus, pumping blood to the rest of the body [161].

AVN cells depolarize automatically only if the SN did not initiate the excitation, or when the excitation does not reach the AVN. Deeper structures like the His bundle, the Tawara branches, and the Purkinje fibers can also depolarize spontaneously if the excitation is not conducted by the AVN [193].

4.5 Cardiac Mechanics

4.5.1 Cardiac Tension Development

The electrical excitation of a myocyte initiates the process called excitation contraction coupling (ECC). The increase of intracellular calcium concentration from the extracellular space triggers the release of much larger amounts of calcium ions stored in the sarcoplasmic reticulum (SR) in a mechanism called the calcium induced calcium release (CICR) mechanism [195]. Due to the high concentration of calcium ions, calcium binds to the filament proteins and enables contraction of sarcomeres in a process that will be described later in this section [196].

As mentioned in 4.3, each muscle fiber is composed of several hundreds of myofibrils. In turn each is a bundle of millions of sarcomeres that constitute the fun-

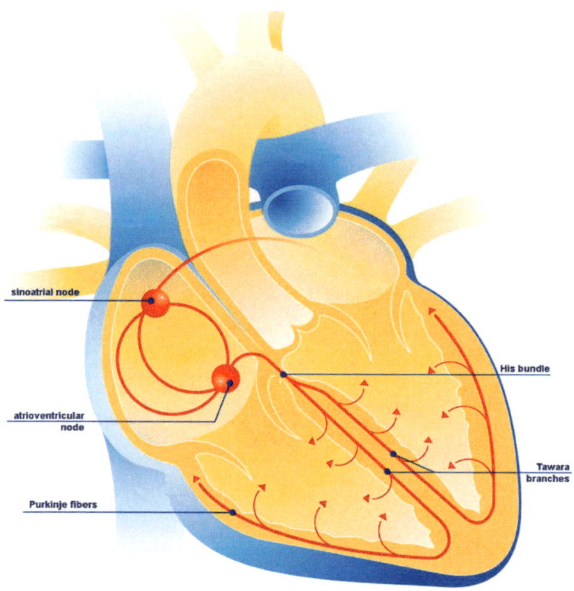

Fig. 4.13 Electrical conduction system of the heart (reproduced from [194]).

damental contractile unit in the cardiac myocyte. The thick filaments are attached to the m-disc in the middle of the sarcomere. Two sets of thin filaments are anchored to the Z disc, whereas the other ends overlap a portion of the thick filament where actin and myosin can interact by building cross bridges. Two successive Z discs define one sarcomere.

Each thick filament is composed of several myosin II molecules, depicted in Figure 4.14(a). Each myosin filament consists of two ball-shaped heads called the motor domain, a neck region and a tail domain which are built of coiled-coil molecules as depicted in Figure 4.14(b). The motor domain contains a binding site for actin and a pocket for the binding and hydrolysis of ATP. Two light chains are located at the neck region, which influences the stiffness of the neck region and the activity of the head group. The tail domains of several hundred myosin molecules are coupled back-to-back to each other forming the thick filament. This means that a thick filament has head groups at both ends, organized in opposite directions. The area where no head groups are located is called the bare zone [2, 197, 184].

Myosin has the ability to change the protein configuration, with the binding and hydrolysis of ATP [184].

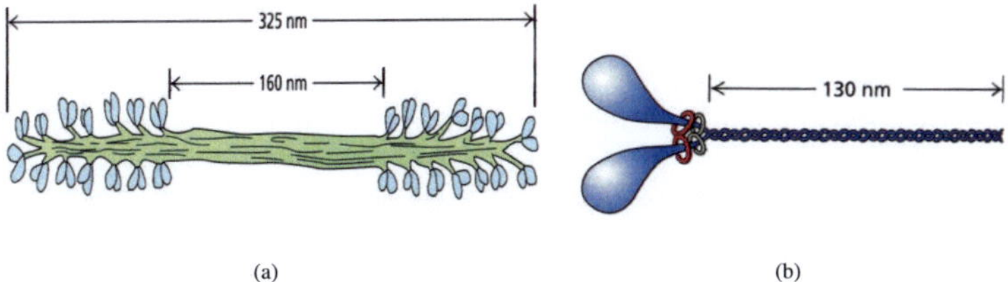

Fig. 4.14 Structure of a thick filament. Several myosin II molecules make up the thick filament (a), zoom in on one myosin II molecule (b) (modified from [184]).

The thin filament is built up of actin, tropomyosin (Tm), and troponin (Tn) with three sub units. The backbone of a thin filament is formed by a double helix of actin stripes that are bound to the Z disc from one side and overlap with the thick filament on the other. A rope-like protein-chain of several tropomyosins, connected to each other head to tail, wraps the double actin helix [198]. In the resting position, the tropomyosin is located between the two actin filaments, hiding myosin binding sites. The troponin complex, composed of troponin C (TnC), troponin I (TnI) and troponin T (TnT), is located at the end of each tropomyosin. TnC is the calcium binding subunit, TnI is the actin bound and actin-myosin ATPase inhibiting part, and TnT the tropomyosin binding site that have a head and a tail part. Figure 4.15 illustrates the thin filament structure.

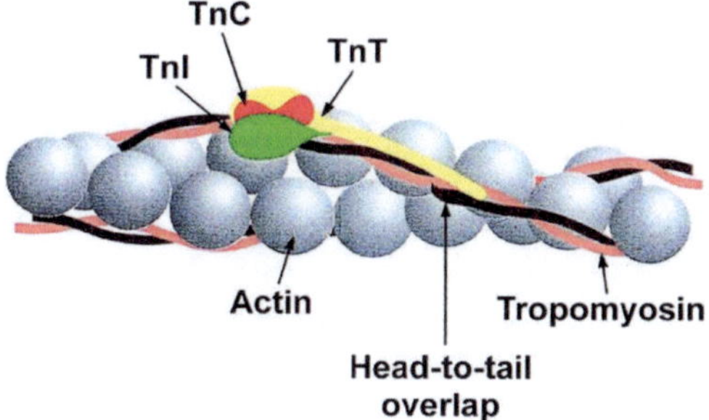

Fig. 4.15 Illustration of a thin filament constructed of three proteins: actin, tropomyosin and troponin (reproduced from [184]).

With the binding of Ca^{2+} to TnC, the troponin-tropomyosin complex is shifted and the, previously hidden, myosin binding site of actin is exposed. This enables the myosin heads to bind to actin, by building cross-bridges. The electrical excitation of the myocyte opens the sarcolemmal voltage-gated L-type calcium channels allowing the diffusion of calcium ions to the cytoplasm. An additional amount of calcium ions get transported to the cell by sodium-calcium exchangers [2]. The increase of the cytoplasmic calcium concentration triggers the calcium sensitive ryanodine receptor to open which results in a further release of calcium from the sarcoplasmic reticulum into the cytoplasm [199].

The cytoplasmic calcium binds to troponin C, causing a shifting of the troponin-tropomyosin complex and thus exposing the myosin binding site. This enables the myosin head to bind to actin in a rigor state [184]. The complex of myosin bound to actin is called cross-bridge [166].

If ATP binds to the myosin head, its binding to actin is dissociated. The binding of ATP results in a change of the myosin protein configuration. This causes the myosin head to move a step towards the positive end of the actin filament where it binds to the next binding site. At the same time the ATP molecule is hydrolyzed to ADP with a Gibbs energy of $-7.3 \ kcal/mol$ [184]:

$$ATP + H_2O \rightarrow ADP + P_i + H^+ \tag{4.4}$$

In a next step phosphate dissociates, and the released energy is used for the conformational change of the myosin head. The myosin head executes the so-called power stroke and moves the actin filament (see Fig. 4.16). After ADP is released to the cytoplasm, the myosin head is restored to the rigor state. The contraction process repeats as long as enough ATP and calcium are available. The process stops if the ends of the thick filaments reach the Z disc [184, 2].

4.5.1.1 Calcium Tension Relationship

The tension resulting of a contraction process in a myocyte is determined by the intracellular calcium concentration. The function relating the resulting tension T to the intracellular calcium concentration $[Ca^{2+}]_i$ can be expressed by the sigmoid Hill equation [184]:

$$T = T_{\max} \frac{[Ca^{2+}]_i^{n_H}}{[Ca^{2+}]_i^{n_H} + [Ca^{2+}]_{50}^{n_H}} \tag{4.5}$$

where $[Ca^{2+}]_{50}^{n_H}$ is the intracellular calcium concentration when the tension reaches the half of its maximum value, and $T_{\max}$ is the maximum tension in saturated calcium state. This is however a very simplified description of a rather complicated process. Since this model was proposed many other models that incorporated new finding in physiology and biology were proposed (see Section 4.7.4).

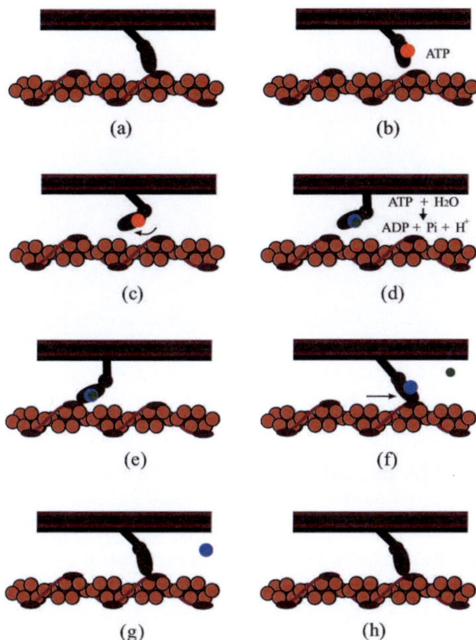

Fig. 4.16 Illustration of the cross-bridge cycle: Tight binding between myosin and actin in the presence of Ca^{2+} (a). Binding of ATP cause a dissociation of the actin myosin binding (b), Myosin head moves due to the change of the protein configuration (c). ATP is hydrolyzed to ADP and phosphate and hydrogen (d). Myosin head binds to the next binding site (e). Myosin head executes power stroke (f). Dissociation of ADP (g). Tight binding between myosin and actin (h) (reproduced with permission from T. Fritz).

4.5.1.2 Stretch Tension Relationship

The maximal tension of the sarcomere depends on the degree of overlap of myosin and actin filaments. If the sarcomere is stretched away off its optimal overlap zone the potentially developed tension drops gradually. Similarly, the more compressed the sarcomere is, the less tension it can generate. The sarcomere has strong parallel elastic structures which counteract an exceeding stretching [184]. The relationship of stretch and tension is shown in Figure 4.17.

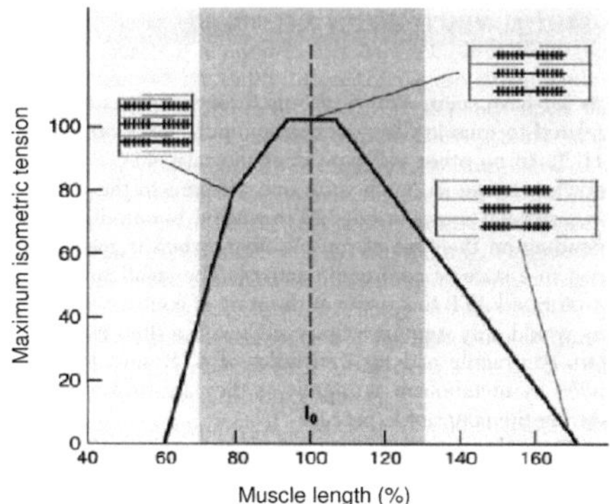

Fig. 4.17 Stretch tension relationship (reproduced from [200]).

4.5.1.3 Cooperative Mechanisms

Generally speaking, the binding and releasing of one protein can influence the binding and release of a neighboring proteins especially if these proteins had active binding sites. Interferences that support the functionality of a protein are called cooperative mechanism. In the case of tension development process, three cooperative mechanisms were reported [2, 184]:

- *Cross-bridge-troponin:* The binding between actin and myosin increases the affinity of the troponin complex for calcium, which supports the building of further cross-bridges.
- *Cross-bridge-cross-bridge:* If the binding between myosin and actin is released, the tropomyosin remains shifted. This allows another myosin head to bind to this binding site.
- *Tropomyosin-tropomyosin:* A tropomyosin protein which changes its configuration due to calcium binding, also influences the probability of its tropomyosin neighbor to release its myosin binding site.

Due to cell stretching or deformation, cross bridges generate a resting tension T_0 in myocytes [184].

4.5.2 Passive Mechanical Properties of The Heart

As mentioned in 4.2, cardiac tissue consists of myocardial muscle fibers arranged in groups. These groups form multiple sheets. A network of extracellular connective tissue, made of collagen, elastin, glyosaminoglycans and glycoproteins, binds the sheets as well as the fibers together [151, 23].

Several experimental studies have been conducted to investigate the mechanical properties of myocardial tissue. Uniaxial, biaxial and triaxial measurement devices were developed and used to measure the stress-strain relationship of myocardial tissue. Table 4.1 shows a list of the most relevant experiments.

Table 4.1 Measurements of mechanical properties of myocardium

Tissue	Species	Date	Publisher
Papillary muscle	cat	1964	Sonnenblick *et al.*[201]
Papillary muscle	rabbit	1973	Pinto *et al.* [202]
Papillary muscle	rat	1974	Janz *et al.* [203]
ventricular muscle	rabbit	1974	Albert *et al.* [204]
ventricular muscle	hamster	1975	Kane *et al.* [205]
ventricular muscle	canine	1976	Rankin *et al.* [206]
ventricular muscle	canine	1988	Hunter *et al.* [207]
ventricular muscle	canine, rat	1991	Guccione *et al.* [208]
ventricular muscle	canine	1994	Novak *et al.* [209]
ventricular muscle	canine	1995	Hunter *et al.* [210]
ventricular muscle	canine	1995	Moulton *et al.*[211]
ventricular muscle	chicken	1997	Miller *et al.*[212]
ventricular muscle	canine	1997	Hunter *et al.* [61]
ventricular myocytes	rat	1998	Omens *et al.* [213]
ventricular muscle	cat	1998	Zile *et al.* [214]
septal muscle	rat	2000	Dokos *et al.* [215]
ventricular muscle	canine	2000	Okamoto *et al.*[216]
ventricular muscle	canine	2003	Dokos *et al.* [217]

The experimental studies showed that myocardial tissue is non-linear, anisotropic and viscoelastic [206]. Furthermore, the change of volume of myocardial tissue undergoing deformation is small enough to consider the myocardium nearly incompressible.

The studies showed that the myocardial tissue has mainly three axis of anisotropy [61, 151]. The first along the fiber direction called the fiber-axis. The second orthogonal to the fiber-axis in the plane of the sheet, called the sheet-axis. And the third normal to the sheet, the sheet-normal-axis, as shown in Figure 4.18.
These axes show different non-linear elastic responses, that can be associated to the organization of cardiac connective tissue mentioned above [23]. Along the fiber axis, the tissue is much stiffer in comparison with the sheet and sheet-normal directions. The stiffness along the fiber can be affiliated to the endomysial collagen surrounding the myocytes and the intracellular titin protein [218, 23]. The stiffness of the sheet axis results from endomysial collagen which connects the fibers of a sheet, while the weak connection of the sheets by perimysial collagen results in

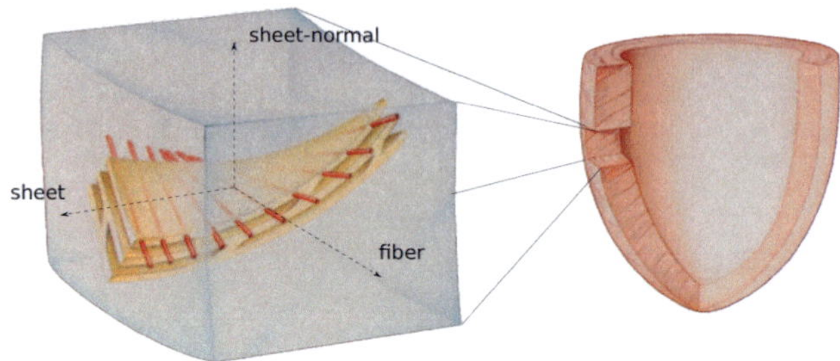

Fig. 4.18 Anisotropy axis, fiber sheet and sheet-normal (reproduced from [153]).

the elastic behavior along the sheet-normal axis [152, 23, 153, 219]. Figure 4.19 shows the stress-strain relation for strain along the three directions of anisotropy.

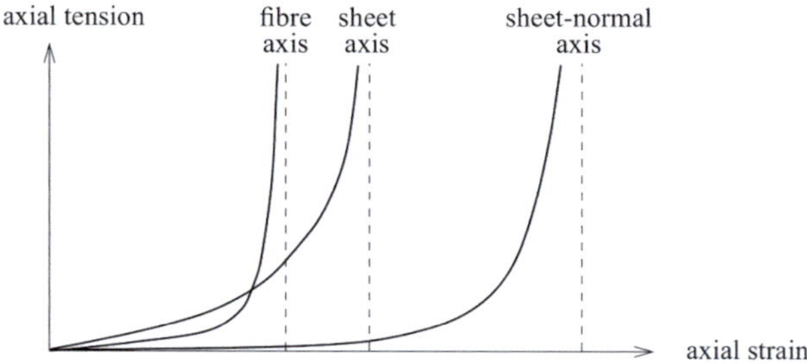

Fig. 4.19 Stress-strain relationship for strain along fiber, sheet and sheet-normal axis of myocardial tissue (reproduced from [23]).

4.5.3 Residual Stress

Residual stress in an organ is the stress that remains when all external loads are removed [220]. Omens *et al.* showed the presence of residual stress in the passive intact heart by cutting an equatorial cross-sectiontional ring from an isolated heart of a rat. As soon as the ring was cut it sprang open into an arc releaving the residual stress [220, 221]. Many other studies support these findings [222, 223, 224].

To better understand the stress state of the myocardial tissue, Omens *et al.* measured the opening angles for $2 - 3mm$-thick equatorial slices of fresh potassium-arrested rat ventricles [220] as demonstrated in Figure 4.20. The opening angles mean was $45° \pm 10° (STD)$ indicating a significant amount of residual stress.

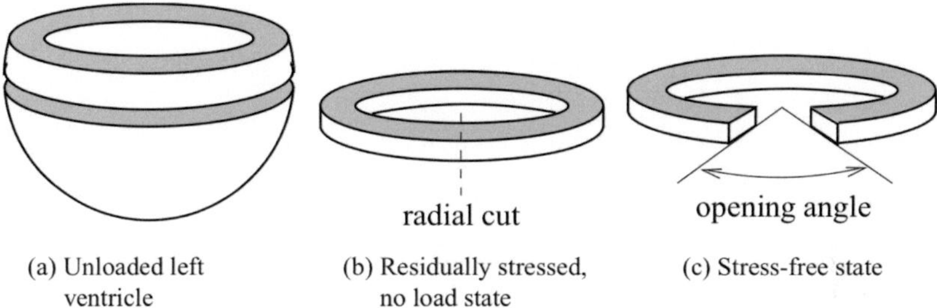

(a) Unloaded left ventricle

(b) Residually stressed, no load state

(c) Stress-free state

Fig. 4.20 Schematic representation of an equatorial slice from an unloaded left ventricle (a), radial cut of the slice (b), the opening angle of the stress-free state (c) (reproduced from [153]).

By tracing the movement of markers placed on the slices surface, it was revealed that the endocardial myocardium of the intact ventricle was under compression stress, and that the epicardial myocardium was under a tensile stress.

Costa *et al.* measured the deformations of the left ventricular wall when stress was relieved locally, and by doing that obtained a three-dimensional characterization of the residual stress that revealed substantial three-dimensional stress components not described previously in the one- or two-dimensional experiments [224].

The analysis showed also that the residual stress-bearing structures tend to align with the local myocardial sheet orientation.

4.5.4 Viscous Damping

T.S. Harris *et al.* conducted experiments to evaluate viscous damping of myocardial tissue [225]. Hereby a patch of canine myocardial tissue was stretched with different velocities and the associated forces were recorded. An exponential function was fitted to the experimental data (see Fig 4.21).

Anisotropic damping properties and shear damping were not considered. Therefore, the resulting function provided an incomplete description of damping in the myocardial tissue. Further experimental data is needed for a better understanding of myocardial damping behavior.

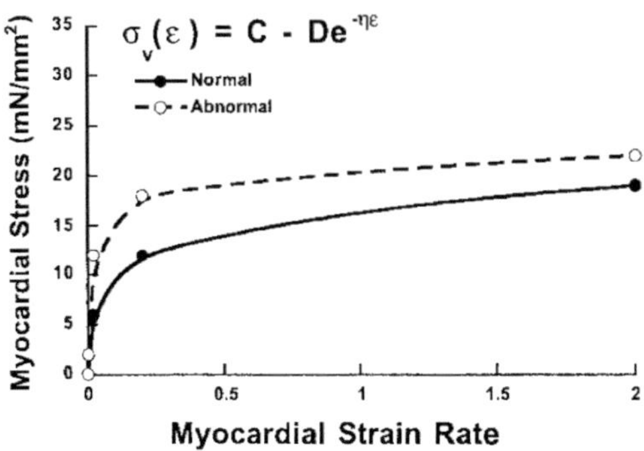

Fig. 4.21 Damping behavior of canine myocardium (reproduced from [225]).

4.6 Cardiac Cycle

The main function of the heart is to pump blood throughout the body delivering nutrients including oxygen and removing wastes including carbon dioxide. To pump blood, the heart beats in a rhythmic manner which is maintained by electrical signals generated within the heart itself.

The cardiac events that occur from the beginning of one heartbeat to the beginning of the next are called the cardiac cycle [160]. Each normal cardiac cycle is initiated by spontaneous generation of an action potential in the SN that propagates throughout the heart as described in Section 4.4.4.

The cardiac cycle consists of two phases:

- Ventricular systole during which the ventricles contract and eject blood into the pulmonary and the systemic circulatory system.
- Ventricular diastole during which the ventricles relax and fill with blood

During ventricular systole, large amounts of blood are pumped into the cirulatory system while AV valves are closed. As soon as the ventricular systole is over and ventricular pressures fall to their diastolic values, AV valves open and allow blood to flowing to the ventricles, during atrial diastole. Then the atria contract (atrial systole), pushing more blood to both ventricles.

After $0.1 - 0.2s$ the ventricular systole begins. Ventricles start contracting and the ventricular pressures start rising. As soon as the ventricular pressures exceed the atrial pressures, AV valves close preventing blood from flowing back to the atria. During this period, pressures evolving in both ventricles are not high enough to open the aortic and pulmonary valves, and the contraction does not result in ejecting blood. Therefore this period is called the period of isometric contraction. When ventricular pressures exceed certain thresholds, the aortic and the pulmonary valves open and blood starts pouring out of the ventricles. This period is called the period of ejection.

At the end of ventricular systole, ventricular relaxation begins, allowing intraventricular pressures to fall rapidly. The elevated pressures in the distended large arteries push blood back towards the ventricles, snapping the aortic and pulmonary valves closed. The ventricular volume does not change during this period. For this reason it is called the period of isometric relaxation, during which the intraventricular pressures fall rapidly to the low diastolic levels. And at the end, the AV valves open and a new cycle begins. Figure 4.22 shows the different phases of the cardiac cycle and the key values and parameters of the cycle.

At the end of diastole, the volume of blood in the ventricles is called the end-diastolic volume (EDV), the end diastolic volume of each ventricle reaches $110 - 120ml$. The volume of blood ejected by a ventricle during ventricular systole is called stroke volume (SV) which is around $70 - 75ml$ in healthy adults. The fraction of the end-diastolic volume that is ejected during systole is called the ejection fraction (EF). The maximum ejection fraction is usually around 63% [160, 143]. EF can be given using the stroke volume (SV) and the end-diastolic volume (EDV) with

$$EF = \frac{SV}{EDV} \times 100\% \tag{4.6}$$

During the heart cycle the atrioventricular plane moves towards and away from the apex, this movement is called the AV plane displacement (AVPD).

During ventricular systole and while atrioventricular valves are closed, ventricles contact pulling the AV plane in the direction of the apex. This displacement results in a pumping effect pushing blood to the arteries, at the same time it generates suction in atria pulling blood from the veins and filling the atria during that phase. As soon as the ventricular diastole begins, the AV plane moves away from the apex toward the base, forcing the blood in the atria to flow into the ventricles while the AV valves are open. This mechanism ensure a fast filling of the ventricles [226].

The factors influencing AVPD are not well understood. It is suggested that AVPD is caused by contraction of subendocardial longitudinal myocardial fibers [227]. Studies must be done to support this hypothesis. Nonetheless, Carlsson *et al.* [228] concluded that the largest part of the stroke volume (SV) is generated at the base

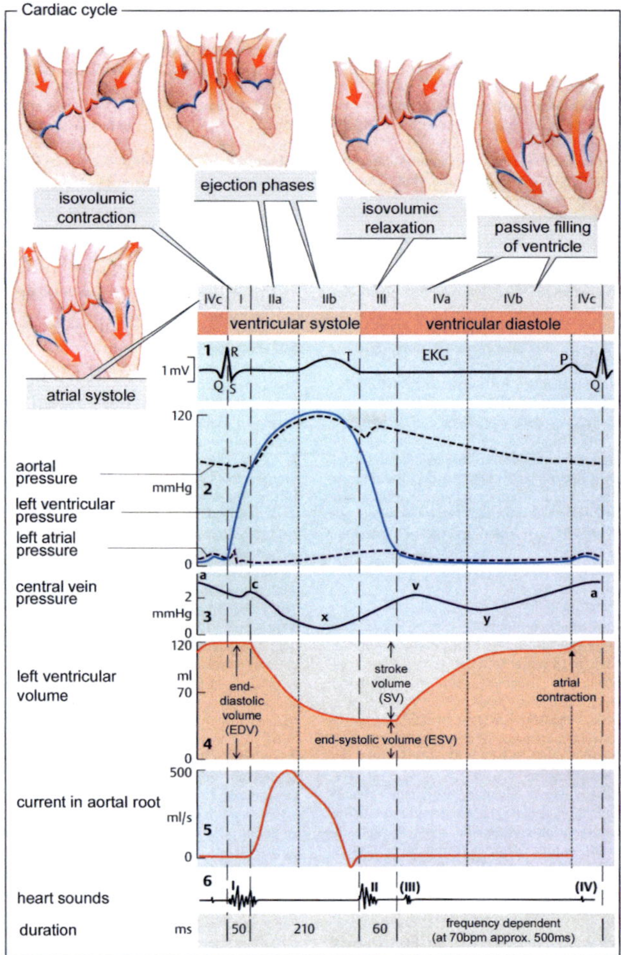

Fig. 4.22 The cardiac cycles and the related parameters (modified from [226]).

of the ventricles and that longitudinal AVPD is the primary contributor to left ventricular pumping, accounting for around 60% of the left ventricle SV.

4.7 Heart Modeling

In this section, methods used for modeling the heart anatomy, physiology and the overall heart function are introduced. The implementation of the modified mass-spring system and the methods used are presented.

4.7.1 Modeling Ventricular Anatomy

Looking back to the complex multi-scale anatomy of the heart, it becomes clear that generating geometry models of the anatomy of heart is a very challenging task. Additional challenges arise from the interest in generating personalized heart models, where inter-individual variations comes into picture.

To tackle these complexities, a set of simplifications and approximations are usually made. In modeling the anatomy of the heart, features that have a relatively high importance for the overall heart function are emphasized, while features of less importance are simply neglected and not taken into account in the resulting models.

Different methods for the modeling of cardiac anatomy have been developed and implemented. Since the study was limited to modeling the deformation of both left and right ventricles, methods for modeling the anatomy of atria were not investigated. In the following, the most relevant methods to this work for modeling the anatomy of ventricles are presented.

4.7.1.1 Analytical Models

To model the ventricles geometry, an analytical description of the left and the right ventricles was developed by McCulloch *et al.* [229, 2]. The left ventricle is described by the crop of two confocal truncated ellipsoids. The ellipsoid focus length d is defined as $d = \sqrt{a^2 - b^2}$ where a is the major radius of the ellipsoid and b the minor radius. A truncation factor $f_\mathrm{d} = 0.5$ is commonly chosen for the truncation of the ellipsoid. f_b is specified by $f_\mathrm{b} = l_\mathrm{be}/l_\mathrm{ea}$ with l_be the distance from the base plane to the equator plane, and l_ea the distance from the equator plane to the apex. The right ventricle is defined by the crop of two confocal truncated ellipsoids with different values for a, b and l_ea. Then the right ventricle model is attached to the left ventricle model by cropping the first with the second.

By using the analytical model to generate a voxel dataset, additional information about fiber orientation and lamination of myocytes can be added to the model as described later in 4.7.1.3

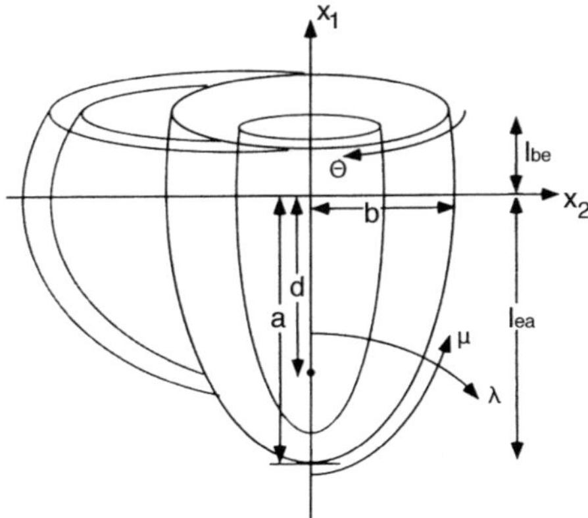

Fig. 4.23 Two truncated confocal ellipsoids representing ventricular geometry, showing, the left ventricle major radius a, minor radius b, focal length d, and spherical coordinates (λ, μ, θ) (reproduced from [229]).

Obviously, this model represents a big simplification when compared to real anatomy of the heart. However, this model proved to be a very helpful asset when performing qualitative experiments.

4.7.1.2 Models Based on Medical Imaging and Image Processing Techniques

3D voxel datasets of the torso or the heart of an individual can be obtained using medical imaging techniques like CT or MRI. These datasets are then segmented and used to generate personalized heart models.

The anatomical models used in this work were derived from MR images of a 27-years-old healthy individual. The images were acquired on a clinical $1.5T$ scanner (Magnetom Avanto, Siemens Medical Systems, Erlangen, Germany).

The cardiac cavities were imaged in a diastolic state with a standard 3D ECG and respiratory gated steady-state free-precession (SSFP, "whole heart approach") sequence with a spatial resolution of $1 \times 1 \times 1 \ mm^3$ [230]. A good differentiation between the blood filled cardiac chambers and the myocardium could be obtained due to the good T1/T2 contrast.

To capture the deformation of the heart an additional Cine dataset was acquired. Hereby, the left and right ventricles were examined from base to apex in the short axis view with a retrospective ECG gated 2D SSFP (trueFISP) Cine (TE/TR: $1.19/35.62ms$, TR-real (echo-spacing): $2.7ms$, : $80°$, slice thickness: $4mm$, in-

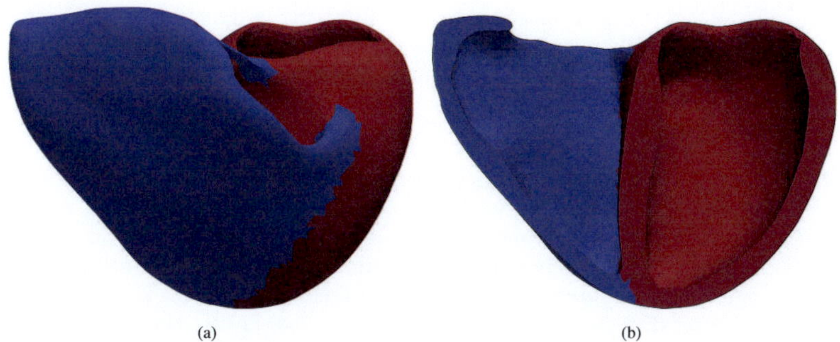

Fig. 4.24 The segmented MRI dataset of left and right ventricles used for the simulations in this work, a front view (a) and a cross section showing the ventricular cavities (b).

plane resolution: $1.9 \times 1.9mm^2$). Using retrospective ECG gating the temporal resolution was $21.6ms$. With RR-Interval $= 1060ms$, 50 phases were generated. Since we are interested in the mechanical deformation of the heart, the first 30 phases (up to $626ms$ after the R peak were segmented.

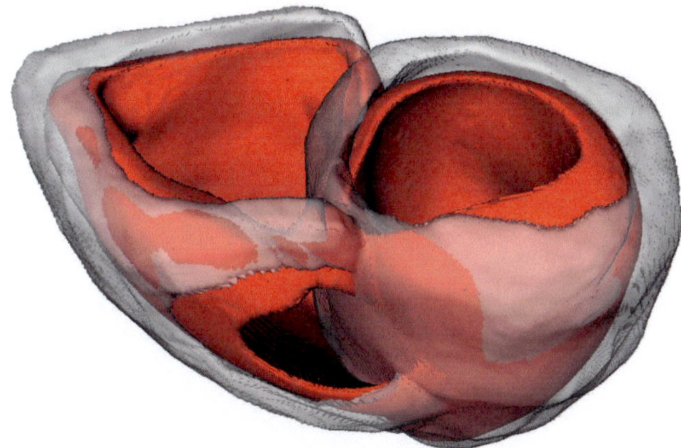

Fig. 4.25 Manually segmented left and right ventricles in diastolic (transparent) and systolic (red) phases of a Cine heart dataset of a 27-years-old healthy individual. In total, 30 phases were segmented.

A segmented Fast Low Angle SHot (FLASH) 2D sequence with prospective ECG gating was also applied. With this technique a $8mm$ grid was "tagged" onto the myocardium (other sequence parameters were: TR/TE: $40.85/3.93ms$; no iPAT, flip angle $14°$, spatial resolution $1.3 \times 1.3 \times 6mm^3$). Overall 22 phases were acquired. These tagging datasets could be used for tracking the myocardial contraction, and are eventually used for the validation and evaluation of modeling methods (see Sections 5.4.1 and 5.4.2). For all these datasets, segmentation was done manually.

4.7.1.3 Modeling Fiber, Sheet and Sheet-Normal Orientations

Available medical imaging techniques are still incapable of extracting the orientation of cardiac myocytes fibers of living individuals. Rule-based approaches based on histological measurements, like the measurements of Streeter *et al.* [155], or reconstruction of diffusion tensor MRI of ex-vivo data [157], can be used to assign fiber orientation to any point of the ventricular myocardium.

In this work, a rule-based method presented by Seemann *et al.* [231] was used to generate the fiber orientation at each point of the ventricular myocardium model. The method was extended to generate the sheet and the sheet-normal orientations, which are important for deformation modeling of the ventricles. The sheet orientation in each point of the model, was set to the orientation of a normal vector on the ventricles surfaces at that point. And the sheet-normal orientation in each point of the model was simply set to the normal on the plane defined by the fiber and sheet orientations.

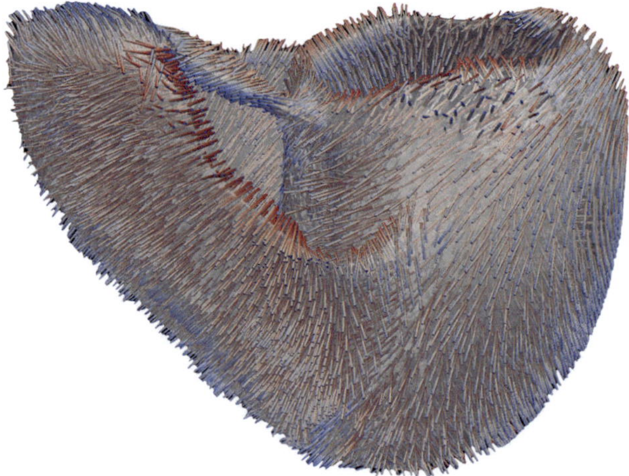

Fig. 4.26 Fiber orientations assigned to the anatomical model of the ventricles using the method of Seemann *et al.* [231]. The size of lines representing the fiber orientation was exaggerated in this visualization.

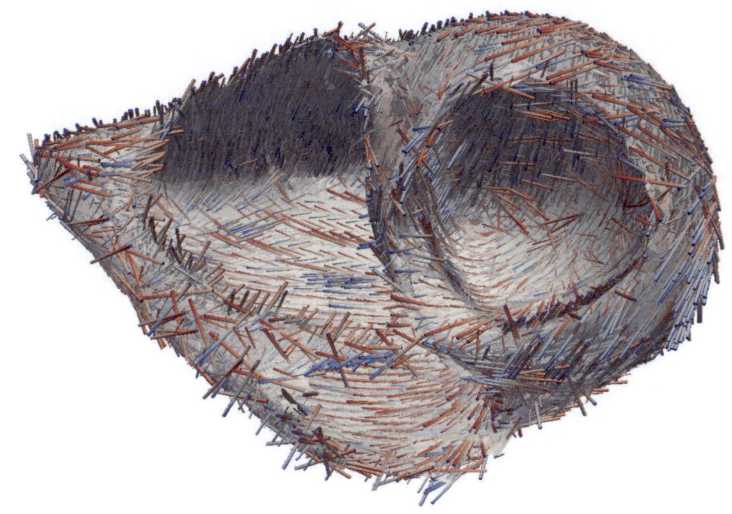

Fig. 4.27 Sheet orientations assigned to the anatomical model of the ventricles. The size of lines representing the fiber orientation was exaggerated.

4.7.1.4 Modeling Anatomical Boundary Conditions

As mentioned in 4.2, the heart is enclosed by the pericardium, and connected to the veins and arteries. These boundary conditions play a significant role in the deformation of the heart, as they constrain its movement.

At the apex, ventricles are strongly bound to the pericardium. This boundary condition was modeled by surrounding the apex with a relatively stiff isotropic nearly-incompressible Neo-Hookean material. This setup is shown in Figure 4.28(b). Some particles of this material were tagged for fixation in all $x-$, $y-$ and $z-$directions. The ventricular walls slide smoothly inside the pericardium. Modeling this interaction is a very complicated task involving modeling the pericardium, the sliding interfaces and the liquid between the interfaces which is beyond the scope of this work. Therefore this interaction was not included in the simulations.

The cardiac skeleton which consist of dense tough elastic tissue was modeled by setting a relatively stiff isotropic nearly-incompressible Neo-Hookean material at the AV plane, as shown in Figure 4.28(c).

Atria, veins and arteries which are in direct mechanical contact with the ventricles, play an important role in the ventricular deformation. The passive elastic properties of blood vessels and the contraction of atria must be taken into account not only in complete heart models but also in biomechanical models of ventricles. In this work atria and arteries were not segmented and not included in the models. However, in an attempt to compensate their passive effect, two hollow spherical caps, of myocardial material, representing the passive atria were added to the model. The tip of the caps were tagged for fixation in all $x-$, $y-$ and $z-$directions. Figure 4.28(d) shows the caps, the apex model and the cardiac skeletal model together.

For the simulations the models were scaled down to reduce computation time. Each of the models contained 71124 tetrahedra only for left and right ventricles. Additional tetrahedra were added to model apex fixation, the AVP, and atrial passive mechanical behaviour.

4.7.1.5 Modeling Conduction System

An accurate anatomical model of the excitation initiation and conduction system is essential to produce a realistic model of the electrophysiology of the heart [232]. Various modeling approaches has been developed to generate a realistic conduction system [233, 234, 235, 236, 237, 238, 232] each varying in complexity and accuracy. In this work a model of conduction system that includes a Purkinje network was used to generate a realistic endocardial stimulation profile [238]. The parameters of the model were iteratively modified to adapt a simulated ECG to a clinically acquired ECG recording of the same healthy proband. Generated isochrone maps have been compared to the work from Durrer *et al.* [239]. The resulting Purkinje network is shown in Figure 4.29.

4.7.2 Modeling Cardiac Electrophysiology

Modeling cardiac electrophysiology is a vast research topic, comprising modeling processes on the cellular lever, up to the organ level. Here we will briefly introduce some topics of cardiac electrophysiology modeling relative to this work.

4.7.2.1 Modeling Cellular Electrophysiology

The computational reconstruction of cellular electrophysiology started with the work of Hodgkin and Huxely [177] who created a model describing the dynamic electrophysiology of a giant squid axon membrane based on measurements of the active and passive electrical behavior. The model of Hodgkin and Huxley describes the membrane of the axon by an equivalent circuit with variable resistors, a capacitor and voltage sources.

This model allows the calculation of diffrent currents passing in and out of an axon's membrane and the transmembrane voltage V_m. The temporal derivative of V_m is given with

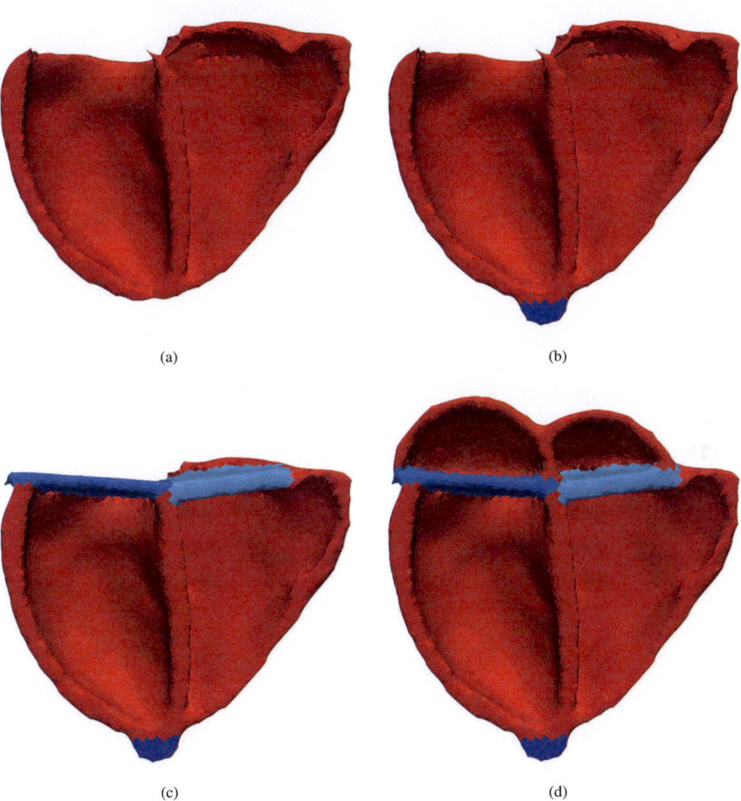

Fig. 4.28 Models used in the simulations, no boundary conditions (a), apex as a sphere of stiff nearly incompressible Neo-Hookean material surrounding the apex and fixed to the frame (b), two cylindric shaped layers of stiff material at the AVP to model the heart skeleton (c), two hollow spheres of isotropic myocardial material fixed at their top to the frame and attached to the AVP added to simulate passive properties of atria (d).

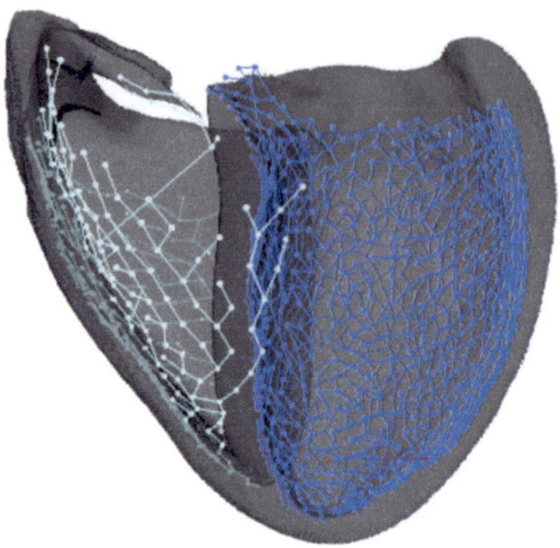

Fig. 4.29 The Purkinje network model generated using the method presented in [238], the method was used to generate a realistic endocardial stimulation profile

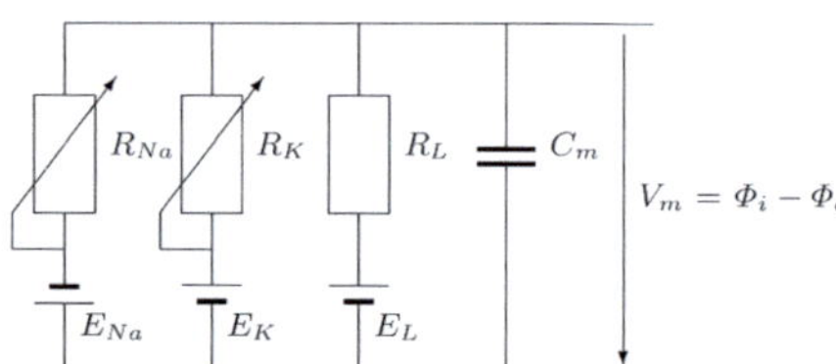

Fig. 4.30 The equivelent circuit of the cell membrane in the Hodgkin Huxley model (reproduced from [184]).

$$\frac{\partial V_{\mathrm{m}}}{\partial t} = -\frac{1}{C_{\mathrm{m}}}(I_{\mathrm{m}} + I_{\mathrm{s}}) \tag{4.7}$$

where C_{m} is called the membrane capacity, I_{s} is the stimulus current and I_{m} is the transmembrane current that can be given by summing the sodium current I_{Na}, the potassium current I_{K} and the leakage current I_{l}:

$$I_{\mathrm{m}} = I_{\mathrm{Na}} + I_{\mathrm{K}} + I_{\mathrm{l}} \tag{4.8}$$

These currents are given by

$$I_{\mathrm{Na}} = g_{\mathrm{Na}}(V_{\mathrm{m}} - E_{\mathrm{Na}}) \tag{4.9}$$

$$I_{\mathrm{K}} = g_{\mathrm{K}}(V_{\mathrm{m}} - E_{\mathrm{K}}) \tag{4.10}$$

$$I_{\mathrm{l}} = g_{\mathrm{l}}(V_{\mathrm{m}} - E_{\mathrm{l}}) \tag{4.11}$$

where g_{Na}, g_{K} and g_{l} are the sodium, potassium and leakage conductances respectively, and E_{Na}, E_{K} and E_{l} are the corresponding equilibrium voltages.

In this model, the ionic concentrations are considered invariant, resulting in a non-varying equilibrium voltages. The leakage current I_{l} summerizes different less dominant ion currents. Here, g_{l} is considered to be constant with the value $g_{\mathrm{L}} = 0.3 \; mS/cm^2$. The remaining conductances are voltage dependent and vary with time. Hodgkin and Huxley used a probabilistic approach to describe these conductances, by introducing gating variables that correspond to states of ion channels. The conductances of sodium and potassium are given by

$$g_{\mathrm{Na}} = m^3 h \breve{g}_{\mathrm{Na}} \tag{4.12}$$

$$g_{\mathrm{K}} = n^4 \breve{g}_{\mathrm{K}} \tag{4.13}$$

where $\breve{g}_{\mathrm{Na}} = 120 \; mS/cm^2$ is the maximum sodium conductance and $\breve{g}_{\mathrm{K}} = 36 \; mS/cm^2$ is the maximum potassium conductance. m and n are called activation gating variables and h the inactivation gating variable. They are time dependent and they describe the state of the corresponding ion channel.

In this model, ion channels can be in an open, closed or inactive state. The transition between two states is described by the forward rate α and the backward rate β that express the transition probabilities between these states. These rates are in turn functions of the transmembrane voltage V_{m}. The dynamics of gating variables can be described by the following differential equations [184]:

$$\frac{dm}{dt} = \alpha_{\mathrm{m}}(1 - m) - \beta_{\mathrm{m}}m \tag{4.14}$$

$$\frac{dh}{dt} = \alpha_{\mathrm{h}}(1 - h) - \beta_{\mathrm{h}}h \tag{4.15}$$

$$\frac{dn}{dt} = \alpha_{\mathrm{n}}(1 - n) - \beta_{\mathrm{n}}n \tag{4.16}$$

The model of Hodgkin and Huxeley presents the base for many mathematical models of membrane electrophysiology dynamics of different cell types. These models extend and adapt the original Hodgkin and Huxeley model to fit the different electrophysiological properties of cell types in question.

Models of cardiac electrophysiology are no exception. Nobel was the first to create a model of Purkinje fibers consisting of four ion channels [240]. With the advances in measurement techniques, cardiac electrophysiology models are constantly en-

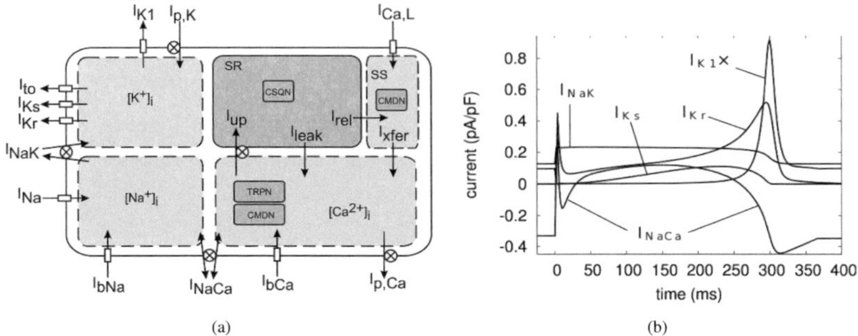

Fig. 4.31 Schematic description of the model of ten-Tusscher II (a) and the ion currents (b). (reproduced from [184]).

hanced and adapted to fit new findings. These models include more ion channels and hence more gating variables related to these ion channels. In this work the model of ten-Tusscher *et al.* was used [241].

The model of ten-Tusscher *et al.* includes a detailed description of human ventricular electrophysiology adapted to experimental data of the major ion currents in human ventricular myocytes. Furthermore it includes an extensive description of the intracellular calcium handling. A subsarcolemmal space, describing the calcium-induced calcium release with a Markov chain for the ryanodine receptor, and the fast and slow voltage inactivation for the L-type calcium current, was also included in the model.

The ten-Tusscher model can reproduce the human subepicardial, subendocardial and M cell membrane currents. Additionally it provides a simplified description of the calcium dynamics [184]. The ten-Tusscher II model provides a more advanced description of the calcium dynamics (see Fig. 4.31(a)) [242]. The total transmembrane current I_m can be given as a sum of all ionic current:

$$\begin{aligned} I_\mathrm{m} =& I_\mathrm{Na} + I_\mathrm{K1} + I_\mathrm{to} + I_\mathrm{Kr} + I_\mathrm{Ks} + I_\mathrm{Ca,L} + I_\mathrm{NaCa} \\ &+ I_\mathrm{NaK} + I_\mathrm{p,Ca} + I_\mathrm{p,K} + I_\mathrm{b,Ca} + I_\mathrm{b,Na} \end{aligned} \tag{4.17}$$

where I_Na is the fast sodium current, $I_\mathrm{Ca,L}$ is L-type calcium current, I_to is called the transient outward current, I_Kr and I_Ks are the rapid and slow delayed rectifier currents, respectively. I_K1 is the inward rectifier current, I_NaCa is the sodium calcium exchanger current, I_NaK is the sodium potassium pump current, $I_\mathrm{p,Ca}$ and $I_\mathrm{p,K}$ are the plateau calcium and the potassium currents and $I_\mathrm{b,Ca}$ and $I_\mathrm{b,K}$, the background calcium and potassium currents. All these currents are in turn described with functions adapted to experimental data [241].

The calcium dynamics model contains a description of intracellular calcium injected through L-type channels and ryanodine receptors. A ryanodine receptor is described by a four state Markov chain, the open state (O), the resting closed state (R), the inactivated closed state (I) and the resting inactivated closed state (RI).

4.7.3 Modeling Excitation Propagation

Several approaches have been developed for modeling the excitation propagation in the myocardium. These approaches differ according to the approximations made in the underlying cellular electrophysiology models, and on the representation of cardiac anatomy. Three main approaches for modeling macroscopic excitation propagation exist: cellular automata [243, 234, 244], reaction diffusion systems [245, 246, 247, 248] and models where electrical currents in the intra- and extracellular spaces as well as the gap junctions are modeled.

In this work, all models of electrical current were used.

Gap junctions can be approximated by a resistor network or modeled by using an intracellular conductivity tensor in a Poisson's equation approach. The conductivity tensor has an advantage because of its ablity to reproduce anisotropic conductivities of heart tissue. The resulting elliptic partial differential equation can be solved by using the finite differences or finite elements method.

Here, a distinction is made between bidomain models and monodomain models.

4.7.3.1 Bidomain Model

The bidomain model consists of two domains representing intra- and extracellular spaces separated by cell membranes. The domains are coupled by the transmembrane voltage V_m. The distribution and density of gap junctions and the conduction properties of the extracellular space are represented in anisotropic conductivity tensors σ_e and σ_i. For each domain, Poisson's equation is defined:

$$\nabla \cdot (\sigma_i \nabla \phi_i) = \beta I_m - I_{si} \tag{4.18}$$
$$\nabla \cdot (\sigma_e \nabla \phi_e) = -\beta I_m - I_{se} \tag{4.19}$$

where indices i, e stand for the intra- and extracellular spaces respectively, I_m is the transmembrane current density, β the surface to volume ratio of a cell, I_{se} and I_{si} are the externally applied current sources. The transmembrane voltage V_m is given by

$$V_m = \phi_i - \phi_e \tag{4.20}$$

and for all areas not containing external or internal stimulus currents, the following equations can be derived:

$$\nabla \cdot (\sigma_i \nabla \phi_i) = -\nabla \cdot (\sigma_e \nabla \phi_e) \tag{4.21}$$

$$\nabla \cdot (\sigma_i \nabla (V_m + \phi_e)) = -\nabla \cdot (\sigma_e \nabla \phi_e) \tag{4.22}$$

and the first part of the bidomain model is then given with

$$\nabla \cdot ((\sigma_i + \sigma_e) \nabla \phi_e) = -\nabla \cdot (\sigma_i \nabla V_m) \tag{4.23}$$

where V_m is the transmemrane potential resulting from the cellular electrophysiological model. From Eqs. (4.18 and 4.19) we can write:

$$\nabla \cdot (\sigma_i \nabla V_m) + \nabla \cdot (\sigma_i \nabla \phi_e) = \beta I_m - I_{si} \tag{4.24}$$

where

$$I_m = (C_m \frac{d}{dt} V_m + I_{mem}) \tag{4.25}$$

results from the electrophysiological models. Finally the second part of the bidomain equations is given by

$$\nabla \cdot (\sigma_i \nabla V_m) + \nabla \cdot (\sigma_i \nabla \phi_e) = \beta(C_m \frac{d}{dt} V_m + I_{mem}) - I_{si} \tag{4.26}$$

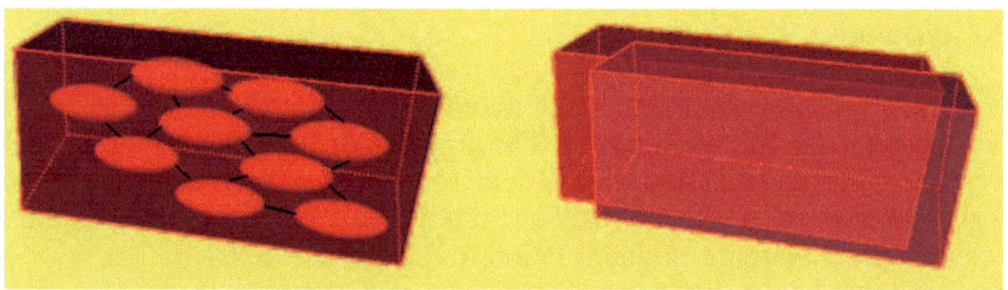

Fig. 4.32 Separated intracellular and extracellular regions (left). Continuous intracellular and extracellular regions (right). (reproduced from [184]).

4.7.3.2 Monodomain Model

By assuming that the conductivity tensors of intra- and extracellular spaces differ only by a scalar coefficient κ, the bidomain model can be transformed into a monodomain model. Hereby, the bidomain Eq. (4.26) can be simplified to [184]:

$$\nabla \cdot (\sigma_i \nabla V_m) = (\kappa + 1)\beta(C_m \frac{d}{dt}V_m + I_{mem}) - I_{si} \qquad (4.27)$$

In this case, just one differential equation has to be solved for V_m. This simplification reduces computational costs.

4.7.4 Modeling Cardiac Tension Development

Mathematical modeling of tension development in muscles dates back to 1938 when Hill *et al.* [249] examined the relationship between heat production and the contraction speed of a frog's skeletal muscle for different loads [184]. In his work, Hill considered only macroscopic information and neglected biophysiological aspects [250]. A.F Huxley, N. Niedergerke, H.E. Huxley and J. Janson developed the sliding filament theory in 1953 [184]. A.F. Huxley presented then a model of the skeletal muscle based on this theory [251]. Since then, many different models that consider or emphasize a variety of physiological aspects have been presented. Many models based on new finding in cellular biology and other related fields of research, have been presented. Table 4.2 lists a number of publications concerning modeling tension development.

Table 4.2 Mathematical models of tension development

Tissue	Species	Date	Publisher
Skeletal muscle	Frog	1938	Hill [249]
Skeletal muscle	-	1957	Huxley [251]
Papillary muscle	mammalian	1980	Panerai [252]
Papillary muscle	rabbit	1991	Peterson, Hunter, Berman [253]
Skinned cardiac myocyte	-	1994	Landesberg, Siedman [254]
Cardiac muscle	-	1994	Landesberg, Siedman [255]
Cardiac muscle	mammalian	1997	Hunter, Nash, Sands [61]
Cardiac myocyte	-	1998	Hunter, McCulloch, ter Keurs [256]
Cardiac myocyte	-	1998	Guccione, Motabarzadeh, Zahalak [257]
Papillary muscle	rabbit	1999	Rice, Winslow, Hunter [258]
Cardiac myocyte	ferret	2000	Rice, Jafri, Winslow [259]
Cardiac myocyte	-	2001	Mlcek, Neumann, Kittnar, Novak[260]
Cardiac myocyte	-	2001	Nickerson, Smith, Hunter [261]
Cardiac myocyte	-	2002	Glänzel, Sachse, Seemann [262, 263]

The hybrid model of Glänzel, Sachse and Seemann [262, 263] was used in this work to model tension development. This model consists of three Markov chains and 14 state variables illustrated in Figure 4.33.

The first Markov chain describes the binding of calcium to troponin with two state variables, the fraction of troponin with no bound calcium (T) and the fraction of troponin with bound calcium (TCa). The time evolution of the binding process can be described with two coupled differential equations:

$$\frac{d}{dt}\begin{pmatrix} T \\ TCa \end{pmatrix} = \begin{pmatrix} -k_{\mathrm{on}} & k_{\mathrm{off}} \\ k_{\mathrm{on}} & -k_{\mathrm{off}} \end{pmatrix} \begin{pmatrix} T \\ TCa \end{pmatrix} \tag{4.28}$$

where k_{on} and k_{off} are the transition probabilities. k_{on} depends on intracellular calcium concentration, the number of cross-bridges (XB) and the stretch of the sarcomere (λ).

The second Markov chain for the tropomyosin configuration is also a two states chain, the fraction of tropomyosin which inhibits the binding of myosin and actin (TM_{off}) and the fraction of tropomyosin that is shifted while the binding site is released (TM_{on}). The differential equation system describing this process is given by

$$\frac{d}{dt}\begin{pmatrix} TM_{\mathrm{on}} \\ TM_{\mathrm{off}} \end{pmatrix} = \begin{pmatrix} -tm_{\mathrm{on}} & tm_{\mathrm{off}} \\ tm_{\mathrm{on}} & -tm_{\mathrm{off}} \end{pmatrix} \begin{pmatrix} TM_{\mathrm{on}} \\ TM_{\mathrm{off}} \end{pmatrix} \tag{4.29}$$

where transition coefficients tm_{on} and tm_{off} depend on TCa and λ. The coefficients of the first two Markov chains incorporate the cooperative mechanism introduced in 4.5.1.3 [184].

The third Markov chain describes the cross-bridge cycle using ten states that consider all possible binding states of myosin, actin, ATP, ADP and phosphate and the different states of the ATP hydrolysis process [184]. The interaction of state variables is also given with a system of coupled differential equations:

$$\frac{d}{dt}\begin{pmatrix} TXB_1 \\ TXB_2 \\ \vdots \\ TXB_{10} \end{pmatrix} = \mathbf{M} \begin{pmatrix} TXB_1 \\ TXB_2 \\ \vdots \\ TXB_{10} \end{pmatrix} \tag{4.30}$$

where TXB_1 to TXB_{10} represent the different state variables of the cross-bridge cycle and $\mathbf{M}$ is a 10×10 matrix that consists of the transition coefficients. These coefficients depend on the stretch and stretch velocity of the sarcomere, and on the amount of ATP, the number of cross-bridges and on the state variable TM_{on} [184, 262].

This model delivers a normalized time evolution of tension T. The normalized tension is scaled with a factor α depending on species and the type of myocardial tissue simulated:

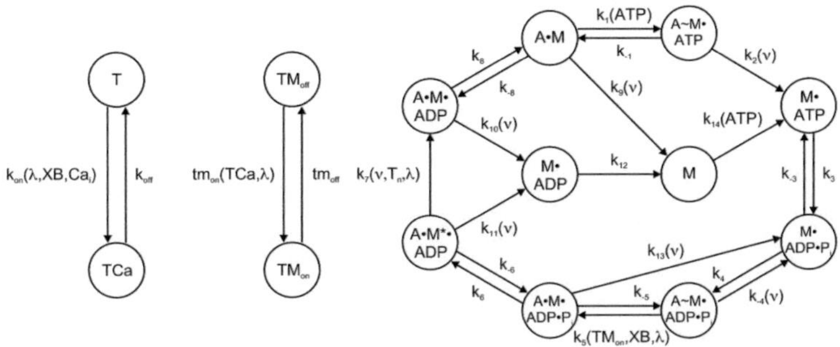

Fig. 4.33 Three Markov chains of the hybrid model. The first describes the binding of calcium to troponin (left), the second describes the configuration of tropomyosin (middle), and the third describes the cross-bridge cycle (right). In the third Markov chain A represents actin, M represents Myosin and P_i represents phosphate with the symbols $\bullet$ for strong and $\sim$ for weak binding. The transition $A \bullet M^* \bullet ADP$ to $A \bullet M \bullet ADP$ represents the stretch dependent irreversible isomerization. The transition $A \bullet M$ to M represent the break of actin and myosin, which is unlikely to occur under physiological conditions (reproduced from [184]).

$$T_\alpha = \alpha T \tag{4.31}$$

In this work, $\alpha = 29.7 KPa$ was used to scale the normalized tension of myocardial tissue [264].

4.7.5 Modeling Elastomechanics of the Ventricles

Modeling cardiac elastomechanics gained a lot of interest in the last two decades driven by advances in physics, biology and the exponential increase of computational power. Around the world, research groups presented different methods to model the deformation of cardiac tissue. Most of these methods use the finite element method (FEM) to solve the partial differential equations (PDEs) provided by continuum mechanics and the theory of elasticity [265, 266, 267, 268, 61, 269, 270, 140, 271, 24]. Hereby, higher order elements [266, 267, 268, 24] or tetrahedral elements [269, 271] were used. For time integration, Euler's methods, the Newmark-Beta method and the Houbolt semi-implicit scheme are often used.

Mohr *et al.* used a mass-spring system in combination with the FEM for modeling the elastomechanics of the heart [272, 273].

In this work, the modified mass-spring system (Adamss) presented in Chapter 2 was used to simulate the passive and active elastomechanics of the ventricular myocardium.

Two different constitutive laws that describe the passive mechanical properties of myocardial tissue were implemented and used. The first is the constitutive law proposed by Guccione *et al.* [274]:

$$W = C(e^Q - 1) \tag{4.32}$$

$$
\begin{aligned}
Q = &\, b_1 E_{11}^2 \\
&+ b_2(E_{22}^2 + E_{33}^2 + E_{23}^2 + E_{32}^2) \\
&+ b_3(E_{12}^2 + E_{21}^2 + E_{13}^2 + E_{31}^2)
\end{aligned}
\tag{4.33}
$$

and the second, presented by Hunter *et al.* [61], was given in Eq. (2.153) and is repeated here for clarity:

$$
\begin{aligned}
W = &\, k_{11}\frac{E_{11}^2}{|a_{11} - E_{11}|^{b_{11}}} + k_{22}\frac{E_{22}^2}{|a_{22} - E_{22}|^{b_{22}}} + k_{33}\frac{E_{33}^2}{|a_{33} - E_{33}|^{b_{33}}} \\
&+ k_{12}\frac{E_{12}^2}{|a_{12} - E_{12}|^{b_{12}}} + k_{13}\frac{E_{13}^2}{|a_{13} - E_{13}|^{b_{13}}} + k_{23}\frac{E_{23}^2}{|a_{23} - E_{23}|^{b_{23}}}
\end{aligned}
$$

E_{ij} are the elements of the Green strain tensor $\mathbf{E}$ and the functions parameters C, b_i, k_{ij} and b_{ij} are obtained by fitting the modeled stress-strain curves to experimental data.

It is clear from Eq. (4.32) that Guccione *et al.* did not make the distinction between the sheet and the sheet-normal anisotropy axes, but rather combined their effect, while the function of Hunter *et al.* reflects the anisotropies along fiber, sheet and sheet-normal axes.

Both functions are just defined for positive values of principle strain. Therefore terms containing negative strains are set to zero [61].

The energy density function of Hunter (Eq. 2.153) is expressed by a rational function with poles a_{ij}. The existence of poles in this function can lead to numerical problems. For instance if the strain E_{ij} became equal to the value of the corresponding pole a_{ij} the function ceases to be defined. If $E_{ij} > a_{ij}$ the resulting stress will have no physiological meaning.

To get around these issues, the terms:

$$k_{ij}\frac{E_{ij}^2}{|a_{ij} - E_{ij}|^{b_{ij}}} \tag{4.34}$$

are defined for the condition $E_{ij} < a_i j - \epsilon$ for a small ϵ. If the condition is not satisfied i.e. $E_{ij} \geq a_i j - \epsilon$, the contribution of E_{ij} to the energy density function W is described by a linear extrapolation of the corresponding term.

Although both methods were implemented, the constitutive law of Guccione (Eq. 4.32) was used in the simulations with parameters adapted to canine myocardial tissue. The parameters are shown in Table 4.3. Parameters for the constitutive law of Hunter *et al.* adapted to canine are given in Table 2.1. Guccione's constitutive

Table 4.3 Parameters for the constitutive law of Guccione *et al.* adapted to canine myocardial tissue [208].

b_1	b_2	b_3	C
15.38	2.0	23.85	1.2

law was also used for the material compensating the passive effect of atria and arteries (see Figures 4.28). Hereby, the parameters b_1, b_2 and b_3 were set to the same value. The values were varied in the various simulations presented in 4.8 to obtain the best AVP mechanism.

The method of virtual hexahedron was used (see Sec. 2.4.2.2) to calculate the active and passive forces in all volume elements of the model.

The scaled tension T_α generated by the tension development model presented in Section 4.7.4 was used to calculate the active forces in the volume elements. The myocardial tension development is a process that generates forces because the count of myofibrils is independent of the deformation. Therefore Eqs. (2.115) must be rewritten after substituting A^t with A^0 as mentioned in Section 2.4.3 to obtain the active forces equations:

$$\mathbf{f}_{2l} = \frac{1}{2} R_v \mathbf{t}_l A_{2l}^0 \hat{\zeta}_l^t \tag{4.35}$$

$$\mathbf{f}_{2l+1} = -\frac{1}{2} R_v \mathbf{t}_l A_{2l+1}^0 \hat{\zeta}_l^t \tag{4.36}$$

where A_{2l}^0 and A_{2l+1}^0 are the surfaces of the virtual hexahedron faces containing intersection points p_{2l} and p_{2l+1} respectively, that axis ζ_l defines computed at time $t = 0$. Since contraction forces act along the fibers in myocardial tissue, ζ_l^t in this case represents the axis in the fiber direction.

Since these constitutive laws do not include volume preservation terms, an extra energy density volume preservation term was added using the method detailed in Section 2.4.2.3. Eqs (2.82) and (2.86) describe the overall energy density function, and the volume preservation term. The parameter p was chosen high enough so that the change of volume of the modeled tissue does not exceed 1%, at the same time, small enough to avoid tetrahedra-locking. A sensitivity analysis was conducted to obtain a suitable value for p. At the end of the analysis the value $p = 5 \times 10^2 kPa$ was chosen and used for all simulations. The Neo-Hookean constitutive law for nearly incompressible material given in Eq. (3.4) was used to model the stiff material surrounding the apex and forming the heart skeleton at the atrioventriclular plane (see Figures 4.28). In the mentioned equation c_2 was set to zero and $\beta = 2c_1 \times 10^5$ was chosen for all simulations to ensure a relatively small change of the volume but not big enough to cause tetrahedra-locking.

Since the electrophysiolgy model and the related tension development model use a regular mesh topology, the resulting tensions are also arranged on a regular grid. To avoid the difficulties of interpolation on unstructured grids, a tetrahedral mesh topology based on a regular grid with six tetrahedra per hexahedron was used for the structure initialization of Adamss (see Section 2.3.1.3). The fact that the segmented dataset used in this work is a voxel dataset is an additional reason in favor of using this topology.

The backward Euler method with adaptive time stepping was used for time integration for all simulations (see Section 2.5.5).

As mentioned in Section 4.5.4, damping in myocardial tissue is not completely understood. Further experimental data that take into account anisotropic damping and shear damping are needed for a better description of damping in myocardial tissue. For this reason the function provided by T.S. Harris *et al.* [225] was not used for modeling damping. Instead linear friction of $2 \times 10^4 N/ms^1$ affecting all particles of the model according to Eq. (2.118) was used in the simulations.

4.7.6 Residual Stress, Resting Stress and the Stress-Free State

In general, constitutive laws consider zero stress at zero strain. As mentioned in Section 4.5.3, Omens *et al.* showed the presence of residual stress in the unloaded myocardial tissue [220], in addition to the resting stress of myocytes at resting length caused by cross-bridges formation [184]. To accurately model the myocardial tissue the stress-free state must be used as the reference state [220, 222, 224, 153].

There is no direct method to measure the residual stress. On the contrary, the strain when a body is deformed from the stress-free state to the no-load state can be quantified. This strain is called the residual strain [220]. Residual strain measurements can be used to calculate residual stress as demonstrated by several research groups [220, 222, 224].

Nash *et al.* approximated the residual strains in his model [23] by introducing the concept of growth tensor presented by Rodriguez *et al.* [222]. The growth tensor, denoted here with $\mathbf{F}_g$, is used to modify the deformation tensor $\mathbf{F}$ to reflect the differences between the no-load state, and the stress-free state of the modeled object [153]. Starting from Eq. (1.3), the modified deformation tensor $\tilde{\mathbf{F}}$ is given with

$$\tilde{F}_{ij} = \frac{\partial x_i}{\partial X_k} F_{g,kj} \tag{4.37}$$

The elements of the growth tensor $F_{g,kj}$ express the deformation gradients, relating the unloaded and stress-free states, with respect to the local material coordinates. By writing the growth tensor as the product of strain and shear strain as in Eq. (2.59), the diagonal elements $F_{g,ii}$ represent the initial extension ratios λ_i^0 due to residual strains for the axis ζ_i:

$$F_{g,ii} = \lambda_i^0 \tag{4.38}$$

the off-diagonal elements represent the initial shear deformation gradients:

$$F_{g,ij} = \lambda_i^0 \gamma_{ij}^0 \tag{4.39}$$

where γ_{ij}^0 is the initial shear strain between axes ζ_i and ζ_j.

Measurements of Omens *et al.* [220] and Rodriguez *et al.* [222] provided values for the initial fibre extension ratio (λ_1^0), whereas neither values for the initial sheet and sheet-normal extension ratios nor for shear deformation gradients were provided. Costa *et al.* [224] quantified distributions of three-dimensional residual strain in the canine mid-anterior left ventricular free wall. This study provided values for the initial fibre extension ratio in consistence with the previously mentioned studies. It provided also values for the initial sheet and sheet-normal extension ratios. Shear components of residual fibre strain were found to be small.

In this work the values of Costa *et al.* were used for the initial extension ratios for fiber, sheet and sheet-normal. While shear components were set to zero. A linear gradient is defined between the endo- and the epimyocardium. Measurements values are set at the endo- and the epimyocardium. A lookup table based on the measurements is used to set values at the midmyocardium. Figure 4.34 shows the lookup table used for setting the residual strain values in the anatomical models. By applying the growth tensor in the unloaded state arises. This stress was used as an approximation for the residual stress needed for the simulations.

To initialize the system with this residual stress, the stress-free configuration must be obtained. An analytical method, presented here, was developed to obtain the stress-free configuration.

The deformation of the model can be described at any moment using the trajectories of all system's particles $\mathbf{u}$, defined in Eq. (2.120), and the initial particle's coordinates vector $\mathbf{u}_{init}$. In the no-load configuration state, at $t = 0$, the trajectories vector will be denoted $\mathbf{u}^0$. And for the stress-free state the trajectories vector will be denoted $\mathbf{u}_{ref}$. The forces vector $\mathbf{F}(\mathbf{u}, \mathbf{u}_{init}, \frac{d}{dt}\mathbf{u}, t)$ defined in Eq. (2.122) can be given at $t = 0$ by

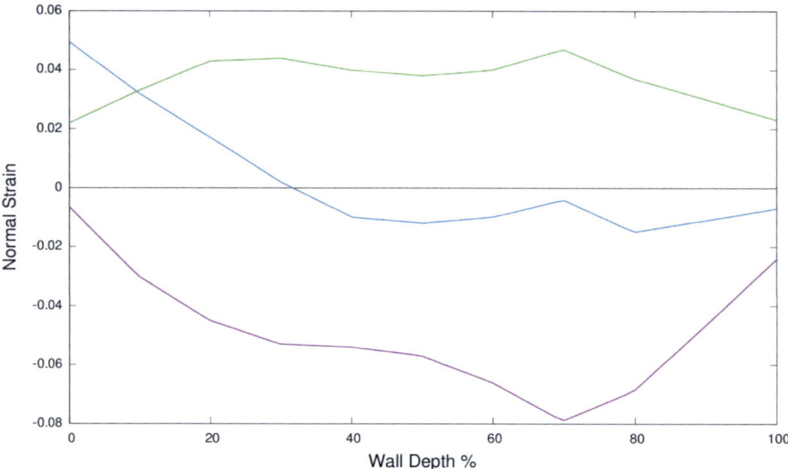

Fig. 4.34 A lookup table for the relation of residual normal strain to the relative wall depth from epicardium (0%) to endocardium (100%) based on the measurements of Costa *et al.* [224] for normal strains in fiber (blue), sheet (green) and sheet normal (violet) directions.

$$\mathbf{F}(\mathbf{u}^0, \mathbf{u}_{\text{init}}, \frac{d}{dt}\mathbf{u}^0, t) \big|_{(t=0)} = \mathbf{F}_\mu(\frac{d}{dt}\mathbf{u}^0) + \mathbf{F}_{\text{d}}(\mathbf{u}^0, \mathbf{u}_{\text{init}}) + \mathbf{F}_{\text{a}}^0 \qquad (4.40)$$

where $\mathbf{F}_\mu$ is the friction force due to initial velocities $\mathbf{v}^0 = \frac{d}{dt}\mathbf{u}^0$. In the case $\mathbf{v}^0 = \mathbf{0}$ this term is eliminated. $\mathbf{F}_{\text{d}}$ is the vector of passive forces resulting from the deformation of the body. These forces depend on $\mathbf{u}$ and $\mathbf{u}_{\text{init}}$. $\mathbf{F}_{\text{a}}^0$ is the initial active forces vector including forces resulting from residual stress.

If residual stress was the only source of active forces, we can write $\mathbf{F}_{\text{a}}^0 = \mathbf{F}_{\text{r}}$ where $\mathbf{F}_{\text{r}}$ is the residual forces vector resulting from the residual stress. By replacing it in Eq. (4.40) we obtain

$$\mathbf{F}(\mathbf{u}^0, \mathbf{u}_{\text{init}}) \big|_{(t=0)} = \mathbf{F}_{\text{d}}(\mathbf{u}^0, \mathbf{u}_{\text{init}}) + \mathbf{F}_{\text{r}} \qquad (4.41)$$

$\mathbf{u}_{\text{init}}$ can be written using the no-load state as

$$\mathbf{u}_{\text{init}} = \mathbf{u}^0 + \Delta\mathbf{u} \qquad (4.42)$$

where $\Delta\mathbf{u}$ is the displacement from the no-load state. By substituting Eq. (4.42) in Eq. (4.43) we get

$$\mathbf{F}(\mathbf{u}^0, \mathbf{u}^0 + \Delta\mathbf{u}) \big|_{(t=0)} = \mathbf{F}_{\text{d}}(\mathbf{u}^0, \mathbf{u}^0 + \Delta\mathbf{u}) + \mathbf{F}_{\text{r}} \qquad (4.43)$$

Finding the reference stress-free state $\mathbf{u}_{\text{ref}}$ is finding the displacement $\Delta\mathbf{u}^0$ such that the no-load state is at equilibrium.

$$\mathbf{F}(\mathbf{u}^0, \mathbf{u}^0 + \Delta\mathbf{u}^0)\,|_{(t=0)} = \mathbf{F}_{\text{d}}(\mathbf{u}^0, \mathbf{u}^0 + \Delta\mathbf{u}^0) + \mathbf{F}_{\text{r}} = 0 \qquad (4.44)$$

Eq. 4.44 states that forces at the no-load state, resulting from the deformation from the reference stress-free state, balance those resulting from residual stress.

By solving the non-linear system in Eq. (4.44) for $\Delta\mathbf{u}^0$ the reference stress-free state is obtained:

$$\mathbf{u}_{\text{ref}} = \mathbf{u}^0 + \Delta\mathbf{u}^0 \qquad (4.45)$$

An implementation of the Newton method (SNES) from the PETSc package [275] was used to solve the non-linear system in Eq. (4.44).

4.7.7 Validation

To evaluate the outcome of simulations, and to compare the different modeling methods and sets of parameters used for modeling, two different validation methods have been employed in this work.

4.7.7.1 Validation with Normalized Volume of Ventricular Cavities

The first method is based on comparing the evolution of the normalized volume $V\%$ of one or both ventricular cavities with measured data, or with values present in the literature [226, 149, 160, 276, 161]. Using $V_{\text{L}(t)}$, $V_{\text{R}(t)}$ or shortly $V(t)$ values and EDV_{ref} at $t = t_{\text{ref}}$ the normalized volume $V(t, t_{\text{ref}})\%$ is calculated using

$$V(t, t_{\text{ref}})\% = \frac{V(t)}{\text{EDV}_{\text{ref}}} \times 100\% = 100\% - \text{EF}(t) \qquad (4.46)$$

To calculate the volume of cavities defined in the model (see 2.3) undergoing deformations, two approaches were implemented.

The first is suitable for mesh topologies based on regular grids. Hereby, volumes of cavity voxels are calculated after deformation by splitting each hexahedron to five tetrahedra, then calculate the volume of these tetrahedra and at the end summing over tetrahedra and eventually hexahedra.

The second approach, is more suitable for mesh topologies based on unstructured grids. In this approach a tetrahedral mesh of the cavity is generated using the nodes or particles that define the surface of that cavity. Hereby the meshing algorithm does not add additional nodes. During deformation, the volume of the cavity can be obtained by summing the volumes of the cavity mesh tetrahedra.

Figure 4.35 shows curves for the normalized volume of the left, the right and both ventricular cavities in one of the modeling simulations. For the volume calculations the first approach was used.

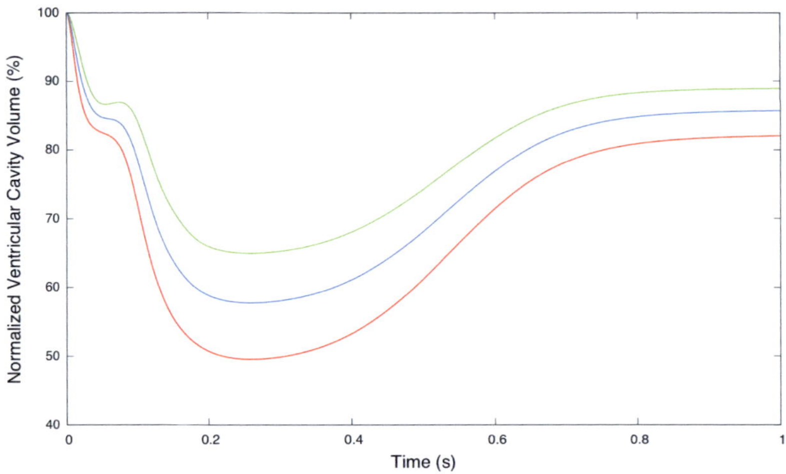

Fig. 4.35 Curves of calculated volume of the left (red), right (green) and both ventricular cavities (blue) each normalized against its respective EDV.

The anatomical model used in this work was derived from MR images as detailed in Section 4.7.1. The additional segmented Cine dataset were used to obtain the normalized volume of the left ventricular cavity V% shown in Figure 4.36. Hereby, the volume of the left and right ventricles, $V_{L(i)}$ and $V_{R(i)}$ respectively, were calculated using the voxel datasets for each phase i of the 30 segmented phases. Since the segmented images are voxel datasets, calculating the volume of a cavity can be easily done by counting the cavity voxels then multiplying by the mesh resolution.

Using $V_{L(i)}$, $V_{R(i)}$ or shortly $V(i)$ values the normalized volume $V(i)\%$ was calculated for all phases using

$$V(i)\% = \frac{V(i)}{\text{EDV}} \times 100\% = 100\% - \text{EF}(i) \qquad (4.47)$$

where the EDV was set to the V(i=1).

$V\%(t, t_{\mathrm{ref}})$ curves resulting from the simulations were validated against the measured curve extracted from the Cine sequence as described above and shown in Figure 4.36.

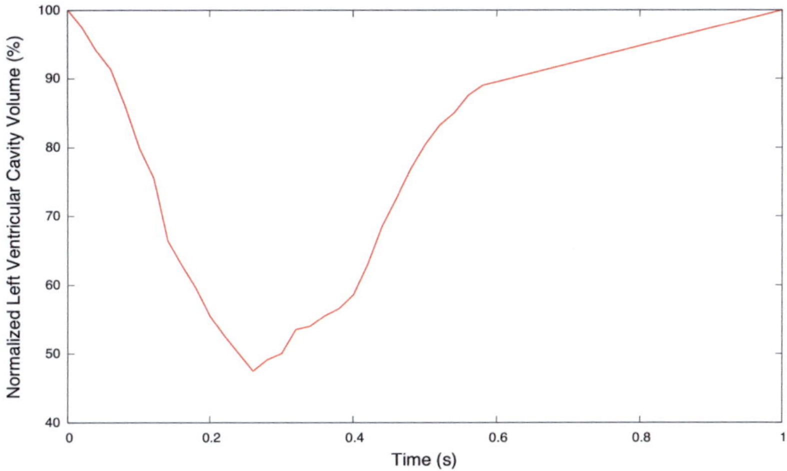

Fig. 4.36 The normalized left ventricular cavity volume V% extracted from a segmented Cine dataset, the same dataset used throughout the heart deformation simulations conducted in this work.

4.7.7.2 Validation Using Elastic 3D Image Registration

Although the V% curves give a good idea about the overall deformation of the models, they cannot be used as a reliable validation measure, because it ignores so many other aspects of the deformation, such as the apex rotation, the atrioventricular plane mechanism and the thickening of ventricular walls. Because validation is very important in the development of any new method, a better validation method that accounts for the mentioned aspects, must be used.

A validation method able to calculate the distance between the nodes of the model and their counterpart in the original data for each timestep of the simulation, is the best case scenario. There are many difficulties and limitations towards this goal. In a medical images dataset, the number of images is always limited, and usually it is small. Therefore validating each timestep of the simulation is not possible. Another major problem is that each image acquisition produces geometries with different number of nodes. Nodes of the model associated with nodes in the original dataset cannot maintain this association in the next timesteps. That means, a direct method to calculate the distances between nodes of the model and the original dataset for different images does not exist.

In many similar applications, markers are used to track the deformation of the studied object. Another method is tracking anatomical features. In both cases, not every point of the model and the original datasets can be compared.

Mapping the nodes of the model onto the nodes of another model, in this case the original medical images dataset, are called image registration.

In this work an elastic 3D image registration method was developed for the validation of heart modeling simulations, and was also used for other applications. This method is detailed in Chapter 5.

4.8 Simulations

For the conducted simulations, the simulation scheme depicted in Figure 4.37 was used. The flow chart shows the overall heart modeling simulation scenario, from the deformation modeling point of view.

Fig. 4.37 Flowchart showing the simulation scheme of the ventricular deformation with Adamss.

Data about tension development in each volume element of the model is needed for each timestep of the ventricles deformation modeling. Since mechanical-electrical coupling was not considered in this work, the data tension development could be generated separately prior to the deformation modeling using the methods described earlier in this chapter.

To model ventricular deformation, the Adamss system is initialized with the geometry data, boundary conditions and the set of parameters specific for the simulation. Then the system enters the deformation calculation loop where deformation forces are first calculated. At this stage, tension development data is introduced to the system. Time integration is then used to update the nodes coordinates which is, in fact, the actual deformation of the model. The system iterates this loop until a predefined simulation duration is over. For all the simulations, the tetrahedral mesh topology with 6 hexahedra in a hexahedron was used for structure initialization, and the method of virtual hexahedron was used to calculate the forces acting on the system's particles. The implicit time integration with adaptive time stepping with a maximal timestep of $10^{-4}s$ was used.

In this set of simulations, the effect of boundary conditions on the ejection fraction, the apex rotation and the AVP mechanism during a full heart cycle was investigated. The effect of blood pressure, valves dynamics, residual and resting stress were not taken into account in this set of simulations.

The investigated boundary conditions included the fixation of the apex, the heart skeleton and the effects of passive elastic properties of atria and arteries represented by the ventricular caps (see Fig. 4.28(d)).

First, the model shown in Figure 4.28(a) was used with no boundary conditions in two full heart cycles simulation. Then, the configuration which includes stiff material surrounding the apex, shown in Figure 4.28(b) was used to investigate the effect a fixed the apex. That was followed by adding the heart skeleton model shown in Figure 4.28(c). After that, several simulations of the model shown in Figure 4.28(d) were conducted to evaluate the effect of the passive elastic properties of atria and arteries on the deformation.

Table 4.4 lists the conducted simulations, the used configurations along with key parameter values.

Figure 4.38 shows the calculated normalized volume of the left ventricular cavity $V_L\%$ of simulations ID: 1, 2, 3 and 4-a plotted with the measured curve. Figure 4.39 shows the calculated $V_L\%$ of simulations ID: 4-a, 4-b, 4-c and 4-d also plotted with the measured curve. And finally Figure 4.40 shows the calculated $V_L\%$ of simulations ID: 1 and 4-d plotted for the full duration of the simulation with the measured curve.

At a first glance, the simulated curves followed the general trend of the measured curve.

However, a large deviation from the measured curve can be seen at the beginning of the heart cycle, where each of the simulated curves are bent in the first $t = 0.1s$. Another large deviation in the $V_L(t, t_{ref} = 0s)\%$ curves was also present after the systole at $t = 0.26s$ and towards the end of the cycle, where the calculated $V_L(t, t_{ref} = 0s)\%$ values increase slower than the corresponding measured values to reach local maxima at $t = 1s$ smaller than the measured value of 100%.

These local maxima are almost at the same level of bents in curvatures at the beginning of the cycle and have the same cause: It is the fact that resting stress at the beginning of these simulations was not taken into account.

Without considering the stress present at $t = 0s$, resulting from the tension development model, the models are not in equilibrium at $t = 0s$. Therefore, a sharp decline in $V_L\%$ can be seen at the beginning of these simulations representing

Table 4.4 Conducted simulations and the corresponding boundary conditions and constitutive laws parameters

Simulation ID	Boundary condition	Constitutive law	Constitutive law's parameters
1	-	-	-
2	Apex fixation	Neo-Hookean	$c_1 = 7.5 \times 10^2 kPa$ $\beta = 10^6 kPa$
3	Apex fixation Heart skeleton	Neo-Hookean	$c_1 = 7.5 \times 10^2 kPa$ $\beta = 10^6 kPa$
4-a	Apex fixation Heart skeleton	Neo-Hookean	$c_1 = 7.5 \times 10^2 kPa$ $\beta = 10^6 kPa$
	Ventricules caps	Guccione	$C = 1.2$ $b_1 = b_2 = b_3 = 15.3 kPa$ $p = 5 \times 10^2 kPa$
4-b	Apex fixation Heart skeleton	Neo-Hookean	$c_1 = 7.5 \times 10^2 kPa$ $\beta = 10^6 kPa$
	Ventricules caps	Guccione	$C = 1.2 \times 10^2$ $b_1 = b_2 = b_3 = 15.3 kPa$ $p = 5 \times 10^2 kPa$
4-c	Apex fixation Heart skeleton	Neo-Hookean	$c_1 = 7.5 \times 10^2 kPa$ $\beta = 10^6 kPa$
	Ventricules caps	Guccione	$C = 1.2 \times 10^{-1}$ $b_1 = b_2 = b_3 = 15.3 kPa$ $p = 50 kPa$
4-d	Apex fixation Heart skeleton	Neo-Hookean	$c_1 = 7.5 \times 10^2 kPa$ $\beta = 10^6 kPa$
	Ventricules caps	Guccione	$C = 1.2 \times 10^{-2}$ $b_1 = b_2 = b_3 = 2.0 kPa$ $p = 50 kPa$

a strong contraction, until the system finds a transient equilibrium state (around $t = 0.080s$) where the curves bent. Towards the end of the simulation with decreasing but not vanishing tensions, each model relaxes to its equilibrium state which differs from the state at $t = 0s$.

In Figures 4.38(a,b) the different boundary condition models are compared where the calculated $V_L\%$ of simulations: 1 (green) , 2 (blue), 3 (cyan) and 4-a (magenta) are plotted with the measured curve (red). The $V_L\%$ curve of the model with no boundary conditions (green) showed the largest ejection fraction (EF) among the conducted simulations. The model showed rotation around the apex-base axis due to fiber orientations. However, because the model was free to move in space and because of loss of energy due to friction, the model relaxed to a configuration that deviated largely from the original configuration. By adding the apex fixation (blue), a better relaxation was observed. Nonetheless this configuration resulted in a smaller EF. By further constraining the model using a stiff material for the AVP (cyan), a large EF as well as a good model relaxation could be achieved. The EF was almost identical to the EF in the case of no boundary conditions, while the relaxation of the model was much better than in both previous cases. However, the AVP movement was not similar to the physical case. The simulation 4-a (magenta) that included, in addition to the apex fixation and the stiff AVP, the passive elastic properties of atria and the fixation of atria, showed a very good relaxation behavior, a better AVP movement but a lower EF.

In Figure 4.39(a,b) different atrial passive elastic parameters are compared, where the calculated $V_L\%$ of simulations: 4-a (green), 4-b (blue), 4-c (cyan) and 4-d (magenta) are plotted with the measured curve (red). Starting with the parameters of the simulation ID: 4-a (green), the resulting deformation showed a good AVP movement and good relaxation behavior. In simulation ID: 4-b (blue), the stiffness of atria was increased, as a result the EF increased, but the AVP movement worsen. In simulation ID: 4-c (cyan), the stiffness of atria was decreased in comparison with values used for simulation ID: 4-a. The EF decreased but the AVP movement improved. A further reduction of the stiffness of atria, in simulation ID: 4-d (magenta), resulted in a better EF in comparison with simulation ID: 4-c and yet a good AVP movement.

Figures 4.40(a,b) show the calculated $V_L\%$ of simulations ID: 1 (green) and 4-d (blue) plotted with the measured curve (red) for two heart cycles.

Figures 4.41(a,b) show snapshots of the model used for simulation ID: 4-d taken from the front and from the back. The white wireframe represents the model in diastolic state while the model in systolic state ($260ms$ of the diastole state) is visualized in solid colors. The snapshots shows the thickening of the ventricular walls, specially the septum and the ventricular wall of the left ventricle. The movement of the AVP towards the apex can also be seen. The rotation of the ventricles can be easily seen specially in Figure 4.41(b) where the difference between the systolic configuration and the diastolic configuration is clear to see.

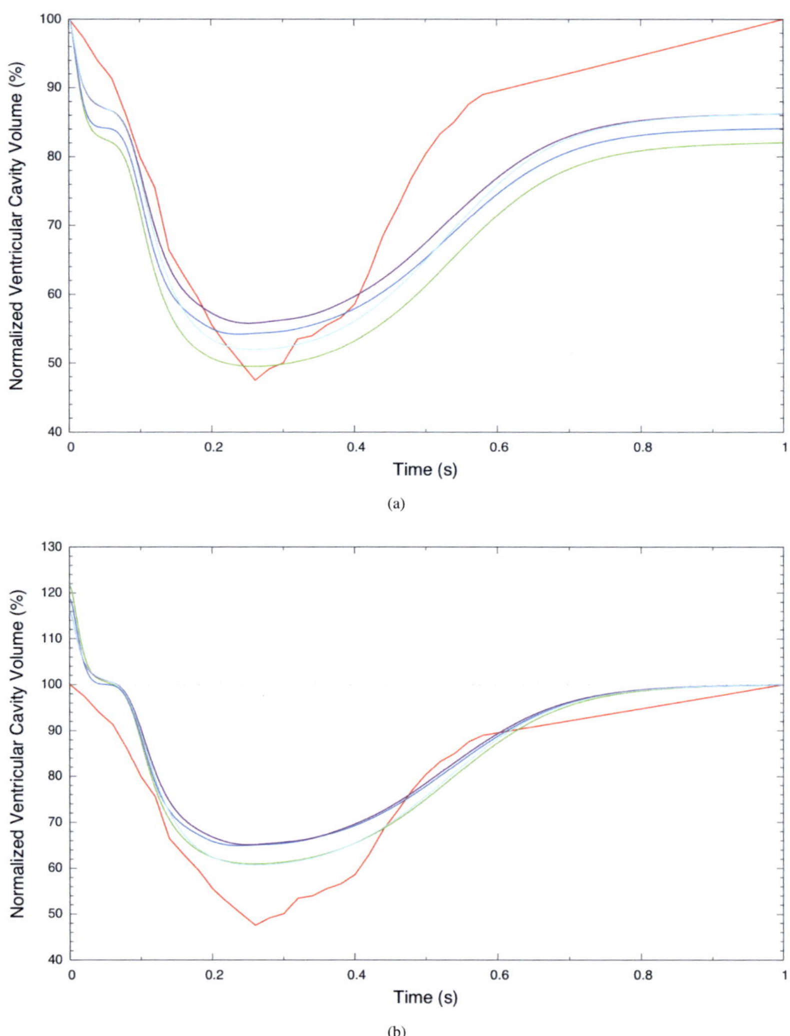

Fig. 4.38 The calculated normalized volume of the left ventricular cavity $V_L\%$ of Simulations ID: 1 (green), 2 (blue), 3 (cyan) and 4-a (magenta) and the measured curve (red). $V_L\%$ calculated with $\mathrm{EDV}(t_{\mathrm{ref}} = 0s)$ (a) and with $\mathrm{EDV}(t_{\mathrm{ref}} = 1s)$ (b).

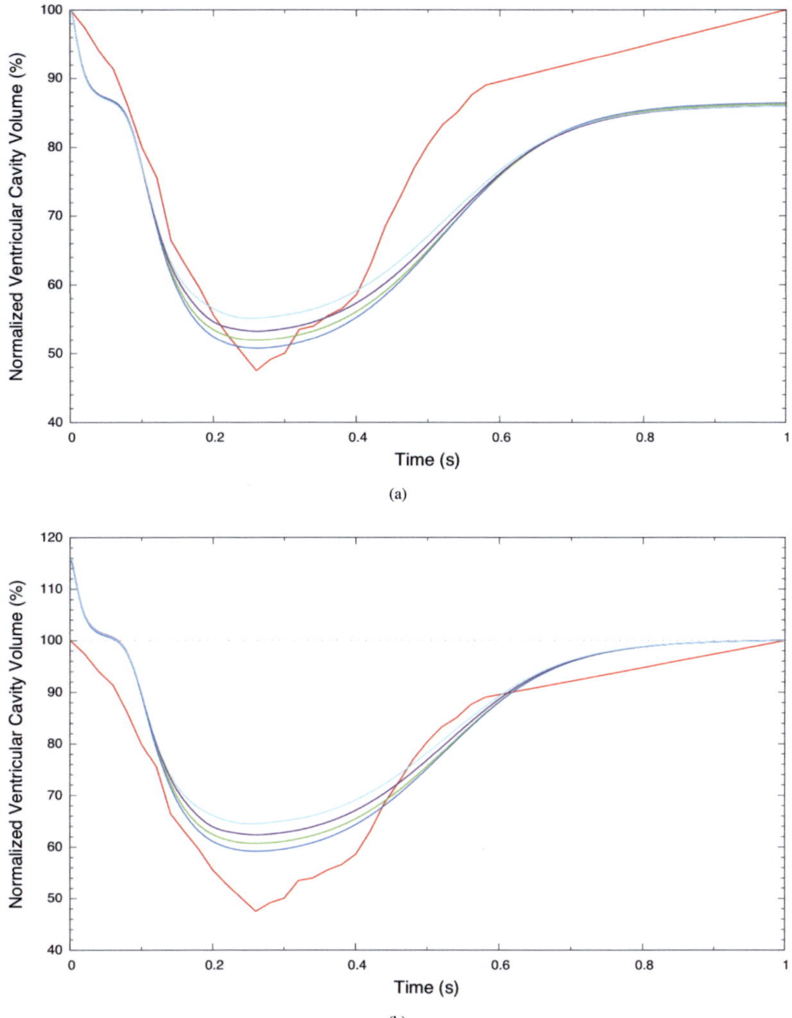

(a)

(b)

Fig. 4.39 The calculated normalized volume of the left ventricular cavity $V_L\%$ of simulations ID: 4-a (green), 4-b (blue), 4-c (cyan) and 4-d (magenta) and the measured curve (red).

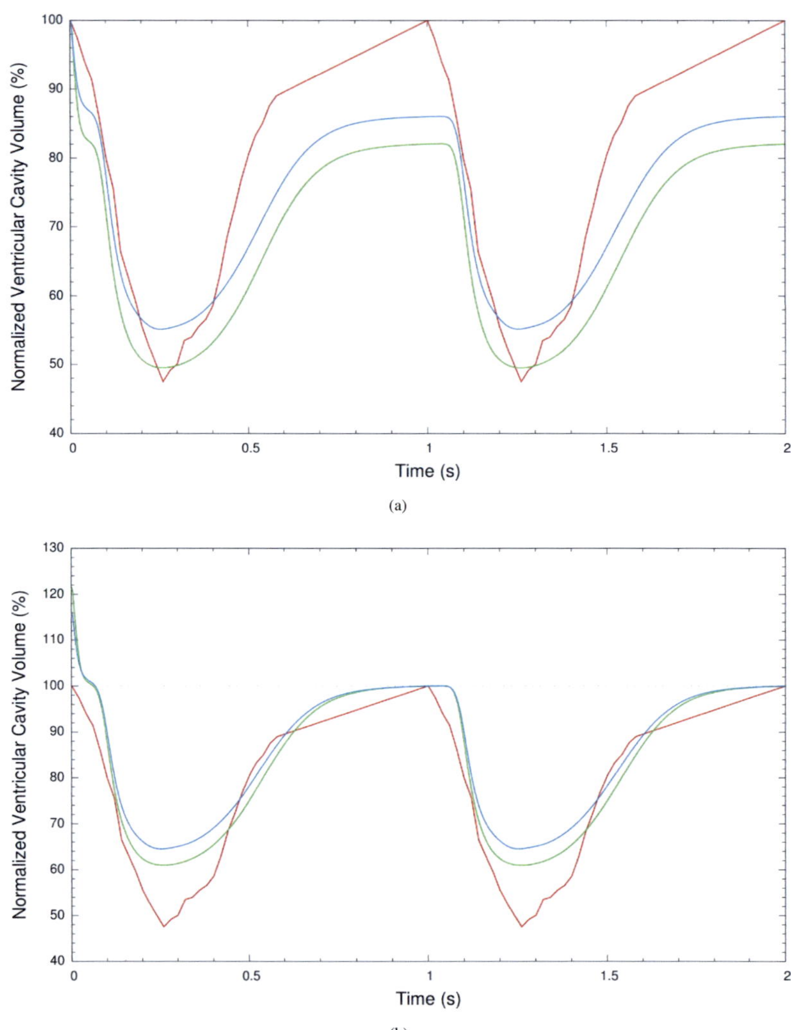

Fig. 4.40 The calculated normalized volume of the left ventricular cavity $V_L\%$ of simulations ID: 1 (green), 4-d (blue), and the measured curve (red). The curves are plotted for two consecutive heart cycles.

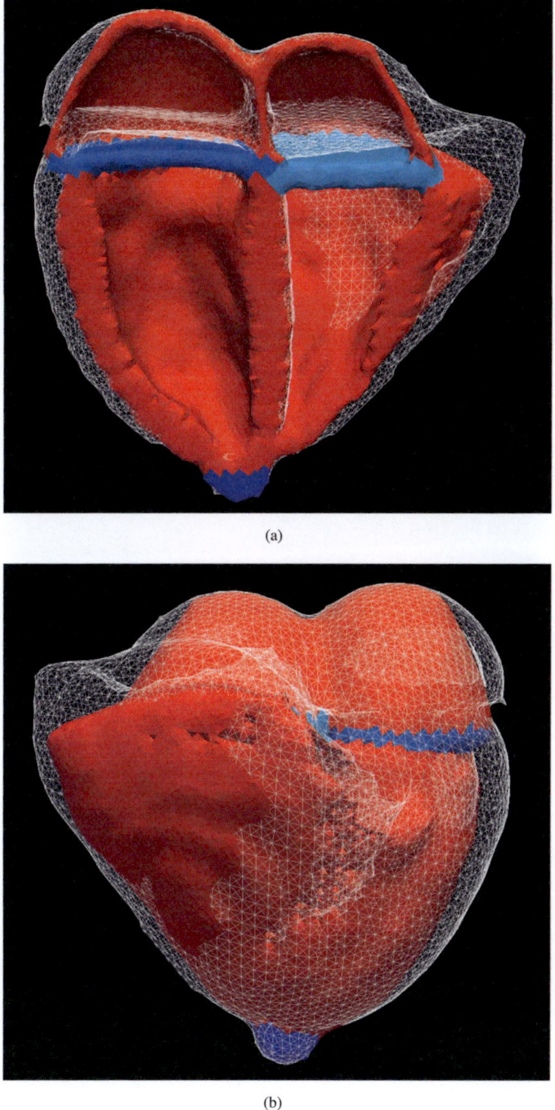

(a)

(b)

Fig. 4.41 Snapshots of the ventricles and boundary conditions model used for simulation ID: 4-d, the white wireframe in both snapshots represents the diastolic configuration while the solid colors represents the deformed model at the systolic state after $260ms$ of the beginning of the simulation.

4.9 Discussion

In this chapter a method for personalized modeling of the passive and active mechanics of the right and left ventricles was presented and described in details in Section 4.7.

Methods used in this work for the generation of personalized anatomical models including a rule-based method to approximate the sheet and sheet-normal orientations were presented in Section 4.7.1. Methods to model apex fixation, the heart skeleton, and the binding to blood vessels were introduced and investigated in Section 4.7.1.4. To obtain the force developed in each volume element of the model, available personalized models of electrophysiology and force development were used in this work. These models were briefly introduced in Section 4.7.2. A detailed implementation of Adamss for the elastomechanical modeling of the ventricles was introduced in Section 4.7.5 including the parameters and settings recommended for simulating the passive and active elastomechanics of ventricles in human. A method for determining the residual stress in the ventricles, and a method for calculating the stress-free state of the model were presented in Section 4.7.6. Finally a set of simulations were conducted and validated with the real imaging datasets in Section 4.8.

The tetrahedral mesh topology was used to initialize the modeling framework of Adamss (see Section 2.3.1.2). As already mentioned, tetrahedral volume elements suffer from tetrahedral locking. Although the effect of locking was mitigated using a nearly incompressible material law with a suitable parameter p (see Eq. 2.86), the effect was not completely eliminated because p could not be chosen too small to keep the maximum volume change of myocardial tissue under 1%.

Generally, higher order volume elements are used to completely avoid this phenomenon [23, 266, 142]. Although no higher order mesh topologies were implemented in this work, the modular design of framework enables the integration of such implementations with ease.

Concerning fiber, sheet and sheet-normal orientations, it should be taken into account that a small number of volume elements across the ventricular walls will result in a bad orientation distribution. The number of volume elements across the ventricular wall must be large enough to reflect a gradual change in, sheet and sheet-normal orientations. Otherwise, the orientations between neighboring volume elements will vary considerably. As a thumb-rule, five volume elements across the left ventricular wall should set the lower limit on spatial resolution.

Papillary muscles and trabecula were neglected in the modeling of ventricles. Because of the shape of these trabecula, during deformation they tend to push against each other. Modeling these muscles requires the implementation of collision detection and collision reaction models. This could adds a heavy costs to the modeling computational needs. According to literature, the volume of these muscles makes less than 1% of the volume of the ventricular cavities [276]. This was verified in the MRI dataset segmented for this work where their volume made 0.718% of the volume of the ventricular cavities. The effect of these muscles on the pumping function of the ventricles is assumed to be very small and negligible. However, a validation of this assumption must be made.

Two different constitutive laws for modeling myocardial tissue were implemented (see Section 4.7.5). Namely the laws of Guccione *et al.* [274] and of Hunter *et al.* [61] (Eq. 4.32 and 2.153). Although the constitutive law of Guccione *et al.* did not make the distinction between the sheet and the sheet-normal anisotropy axes, it was prefered over the constitutive law of Hunter *et al.* because the latter has poles that make the system prone to numerical instability when the strain in each volume element is around poles regions.

The passive mechanical behavior of myocardial tissue was considerably refined in the work of H. Schmid [277]. H. Schmid compared the phenomenological laws of hyperelastic materials based on different strain energy density functions and concluded that the available models have theoretical deficiencies and partially could not fit some deformation modes. Schmid refined the model of Costa *et al.* [268], and evaluated it using a finite element mesh undergoing several non-homogeneous deformation modes. This and other models can be quickly implemented in the Adamss framework. However personalizing these models remains to be a difficult task due to lack of in-vivo techniques to measure or quantify these parameters.

Since damping in myocardial tissue is not completely understood, linear friction of $2 \times 10^4 N/ms^{-1}$ was used. The ventricles rotate around the apex during contraction due to fiber orientation. During relaxation, the ventricles rotates back and swing due to inertia. The value of friction must be chosen to allow for this effect of inertia. This could be done by monitoring the system's total kinetic energy curves where this swing effect can be detected and choosing the friction accordingly.

In this chapter methods to model only the left and right ventricles were presented. However both ventricles are bound to the upper chambers, and they are in direct physical contact with the pericardium, arteries, and of course blood. In order to obtain a more accurate ventricular deformation these important factors should be taken into account.

Modeling atrial elastomechanics can be done using methods very similar to those presented here for the ventricles. The geometry model of atria will differ significantly in that fibers, sheet and sheet-normal orientations follow a completely different distribution in atria than in ventricles [159]. As in modeling ventricles, for better simulation results, the bond with ventricles, and the contact with the pericardium, arteries and blood must also be included in the model when simulating atria.

The pericardium plays a very important role in constraining the deformation of the heart. The chamber walls slide smoothly inside the pericardium that encloses the heart and protects it from over-streching. To reproduce the deformation of the heart accurately, modeling of the pericardium, the sliding interfaces and the lubricant between the interfaces has to be included. In this work, no methods for modeling the mechanics of sliding interfaces were implemented.

Modeling the sliding interfaces involves defining particles belonging to a specific surface, and the implementation of methods for particles surfaces collision detection and reaction. Accurate methods for collision detection and reaction increase the computational complexity of heart modeling and are therefore unsuitable for efficient modeling of the heart.

It is possible to consider the interface between the ventricles and the pericardium to be similar to a very weak elastic bond. This can be done easily with Adamss by surrounding the ventricle by an Neo-Hookean material with small shear and bulk moduli. However the validity of this very simplified model of the pericardium must be verified. The analysis of interaction between the heart muscle and blood in the cavities of the heart, called coupled analysis, is a very important aspect of heart modeling that was not investigated in this work. Coupled analysis includes modeling blood flow in the heart, and the flow in the circulatory system. It also includes modeling blood pressure and modeling the valves of the heart.

The calculation of flow-structure interaction was introduced by Peskin and Mc-Queen [278, 279] where the coupling takes place by describing the fibers of myocardium and valves as discrete elastic fiber filaments embedded in the flow. Vierendeels *et. al* [280] used other flow-structure coupling methods where the coupling takes place via blood pressure. Krittian [281] presented a time-dependent geometrical model of the ventricle and atrium for the calculation of blood flow during a complete cardiac cycle.

Figure 4.22 shows the blood pressure curves in the left ventricle, atria, the aorta and the central vein over a heart cycle. During the ventricular contraction in a normal heart cycle the ventricles pass through four phases: the isovolumic/isovolumetric contraction during which the left ventricle contracts pushing the incompressible blood while the aortic valve is closed. Blood pressure rises while the volume of the cavities remain unchanged until the pressure reaches the aortal diastolic pressure and the aortic valve opens. Due to the isotonic contraction of the left ventricle blood is ejected into the arteries. In this phase blood pressure rises to reach a peak before it starts falling again. When the pressure is low enough the aortic valve closes and the isometric relaxation phase begins where the ventricle relax and intraventricular pressure drops gradually. When the atrial pressure becomes higher than the ventricular pressure the mitral valve opens and the isotonic relaxation phase begins [226].

Modeling blood pressure involves modeling each of the phases and the valves dynamics. This can be done by using a state machine to describe the valves dynamic and making ventricular pressure depend also on the state in which the heart is and on models specific for the blood dynamics of each phase.

Modeling blood pressure in the isovolumic contraction phase can be simply done by imposing a constant volume constraint on the ventricular cavities during deformation. When the pressure reaches the aortal diastolic pressure, the valves state machine switches to the state where the aortic valve is open, and a mathematical model can be used to compute the pressure by coupling the ventricular cavity to a model of the circulatory system. A lumped parameter model based on the so-called Windkessel concept with given boundary conditions can be used. This model adjusts pressure according to the ventricular volume modifications due to mechanical deformations.

This method can be well combined with the coupled analysis model of blood flow of Krittian [281]. During the relaxation phase another model for blood pressure can be used as in the model of Verdonch *et al.* [282] and Weiss *et al.* [283].

In heart modeling, several models describing different phenomena related to the heart must be coupled together. This kind of modeling is called multi-physics modeling where different models are coupled together through defined interfaces. Also, these models could be describing phenomena on the micro scale up to the macro scale in what is called multi-scale modeling.

When designing a computational model, the designer should keep in mind that it could be used in a multi-physics modeling task, and therefore should anticipate the need for providing one or several interfaces for the model.

The different nature, scale and time resolution of used models put various challenges on the design and implementation of the interfaces.

For example, the force development model couples the electrophysiological model of the heart with the model of hearts elastomechanics by passing forces generated due to excitation to the model of elasticity. In the ventricles deformation simulations conducted in this work, the anatomical model used in the electrophysiology simulations had $0.4mm$ resolution whereas the mesh used for modeling elastomechanics had $2mm$ resolution. In this case averaging was used to rescale the resolution of the first model to suite the second. Another issue concerning coupling of these two models is temporal resolution. In each timestep in the elastomechanical model, forces generated using the force development model should be available. Both models must be "synchronized".

However, if the electrophysiology model does not depend on the state and configuration of the model of elastomechanics, the electrophysiology simulations could be conducted first, followed by simulating deformation using the precalculated force development values for every time step of the elastomechanics simulation. Although it is known that deformation affects electrophysiology, the latter was assumed to be independent in this work, and therefore the force development calculations could be conducted separately prior to the deformation simulation.

Since the elastomechanical model uses adaptive time stepping with a minimum limit on reducing the size of the time step. Therefore to provide all possible force development values means to conduct the electrophysiology modeling simulations with this minimum timestep and save the values for each timestep, which is a very time consuming and computationally expensive task. Therefore force development values were calculated with time resolution of $1ms$. Values needed for modeling elastomechanics that were not directly available were linearly interpolated using the nearest upper and lower ms. The temporal and spatial resolution problem occurs frequently when dealing with physical models interfaces.

A method to calculate the reference state of a model at equilibrium under stress was presented. The method was implemented to obtain the reference state of the ventricles under residual and resting stress. This method involves solving a non-linear system (Eq. 4.44). An implementation of the Newton method (SNES) from the PETSc package [275] was used for that purpose. However, the solver did not converge on a solution despite intensive experimentation with the long list of solver parameters.

By trying to solve the system in Eq. (4.44) in an iterative loop where the stress grows over a finite number of steps from zero-stress to the value of residual and resting stress, solutions for 24% of residual and resting stress could be obtained. That is the reference state of the ventricles if the resting and residual stress were 24% of their typical values which is far below the hopes and expectations.

More investigation in the cause of this problem followed by suitable experimentation using numerical methods and techniques should be conducted to solve the non-linear system. This method does not only help in ventricles deformation modeling but also to find the reference state in breast modeling for example.

When calculating residual stress, the sheer strain was not taken into account. A more complete implementation would have to calculate the residual stress for the 3D distributions of residual strain and not only the residual strain in the fiber directions. Costa *et al.* quantified the 3D distributions of residual strain [224] and the results can be used for this task using the same method described in Section 4.7.6.

*****

So many questions surrounding the heart remain unanswered. It is said that "the heart has its reasons", but will it be long before it decides to reveal its mysteries? The keys to the heart lay in the hands of science. But it is still going to take some effort to unlock all its secrets.

Chapter 5
Image Registration with Morphi

5.1 Motivation

In the course of developing the methods used for the mechanical modeling of the heart, the need for a method to validate and compare the resulting deformations arose.

The use of physiological values like the ejection fraction of the ventricles, the maximum angle of apex rotation, the displacement of the atrioventricular plane proved to be helpful to assess the overall deformation, but fails to account for the details of the deformation progress.

In the search for a reliable validation method that addresses these shortcomings, elastic image registration came into the picture.

In this work an elastic 3D image registration method that uses the modified mass-spring system as one of its components, was developed for the validation of heart modeling simulations. The method was also used to track the contraction of the ventricles in a cinematographic and tagged MRI data, and for creating individual human torso models for virtual radiation protection scenarios.

In the following sections, the image registration problem is explained and a mathematical formulation of the problem is laid out. Differences among image registration methods are then discussed, and a classification of these methods based on the work of J. Modersitzki [12], is presented. After that, the focus is set on the Iterative Closest Point algorithm as well as on elastic registration, due to their relevance to this work. Section 5.3 details methods developed and used in this work. Image registration simulation results obtained in this work are presented in Section 5.4 and the methods and the simulation results are discussed at the end of Section 5.5.

5.2 Introduction

In image processing, it is of great interest, not only to analyze single image information, but also information of different images, in order to compare or combine them, or to perform other analyzing tasks using the information. Transforming several image data to the same coordinates system is one of these tasks.

*****

For example, the museum d'Orsay in Paris holds a very valuable collection of art artifacts including paintings, sculptures, furniture, and photography, and an extensive collection of impressionist and post-impressionist masterpieces by painters such as Monet, Manet, and Van Gogh. To assess the deterioration of the paintings, high resolution photos are taken of each of the painting periodically. For this task the best available photography equipments are used. Year after year technological advancement makes it possible to obtain photos with higher resolution, better contrast, and less spatial distortion. For each of the paintings, images taken at different times with different equipment must be aligned properly to allow for comparing the images and thus to assess the effect of time.

In another application, also related to art, sculptures are scanned using 3D scanning systems, and then digitized, in order to make them available for art students and art fans around the world, mainly by making the 3D digital version of these sculptures available online to download. Small sculptures can be placed on a rotating table and scanned directly by a fixed 3D laser scanner. Alternatively the laser scanner can rotate around the sculpture. This method is useful for heavier sculpture. However, the mentioned systems are not suitable for scanning large sculpture like David of Michelangelo at the Galleria dell'Accademia in Florence (Fig. 5.1(a)), or the Great Buddha of Kamakura in Japan (Fig. 5.1(b)), because these systems cannot cover the entire object in one scanning sweep. One solution is to make several range images for the sculpture, after that, these range images are aligned to produce the complete 3D digital version of the sculpture [284, 285].

*****

Also in medicine, this task plays a very important role. For example, in assessing the efficiency of a therapy for a lung cancer patient, by comparing imagine data of his lungs before and after treatment. Or in treatment planning where the location of the cancer has to be accurately determined which is often done by aligning a PET image that shows the location of cancerous cells to a high resolution CT dataset of the patient.

The problem of aligning images as in the examples presented above is called image registration.

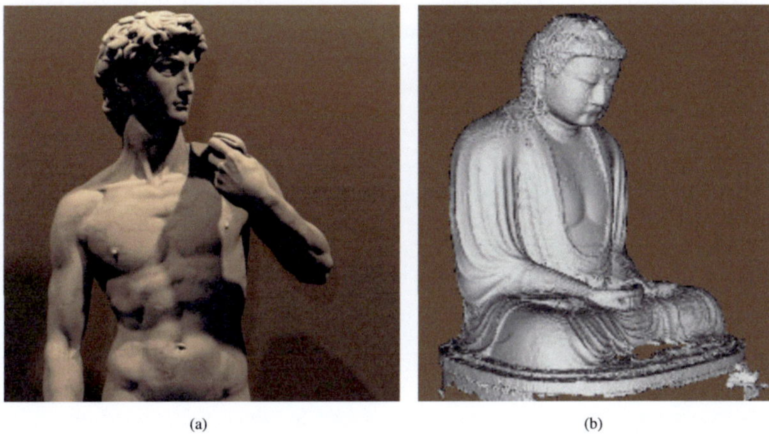

(a)   (b)

Fig. 5.1 3D image registration of sculptures is a difficult image registration task. David of Michelangelo was scanned and registered along with other sculptures as part of the Michelangelo project (modified from [284] (a). The great Buddha of Kamakura proved to be even more challenging due to it's immense scale (modified from [285]) (b).

Definition 5.1. Image Registration is a fundamental tasks within image processing that involves the use of information or data extracted from different images to find an *optimal* geometric *transformation* between *corresponding* image data [12].

It is clear from the examples that the need for image registration recurs in fields ranging from astrophysics, geophysics, computer vision to genetics and biology.

Image registration finds also many applications in medicine. For example, in the analysis of temporal evolution, image registration is used to help physicians perform a diagnosis based on the evolution of tumors, lesions, or anatomical structure through time. therefore aligning medical images taken at different periods is needed [286, 287, 288, 289].

Multimodal images fusion, also called matching or multimodal registration, is another widely spread application of image registration, where the template and reference images are generated using different imaging techniques (like CT, PET, MRI, etc.) to provide physicians with more information about the imaged objects [290, 291, 292, 293, 294, 295, 296].

Registration is also used in inter-patients comparison [297, 298] where the anatomy of function of an organ for example is compared among healthy and un-healthy subjects to gain knowledge about the causes and effects of a specific pathology.

Another medical application where image registration is used, is the construction of 3D volumes. Using a series of continuous 2D slices for example [299, 296]. Sometimes, the acquired images are not complete, in those cases, image registration is used to reconstruct the 3D volumes using the partial data in addition to apriori knowledge [300].

In general, new image registration techniques find their way to medical applications rather slowly, because medical applications demand high accuracy, robustness and reproducibility. Other factors like computational costs and image registration time play a very important role in applications where short registration time is crucial for example in intra-surgical registration.

5.2.1 Mathematical Formulation

Let's start with two images of the same object, in which the imaged object was moved, between acquisitions, relatively to the imaging device. The object in the resulting images will be not aligned spatially, or not registered. This is a typical image registration problem that can also occur if the imagine device was moving between acquisitions, or if the imaged object changed its form between acquisitions, or if different imaging devices were used for the task. Obviously, combinations of these incidents as well as other incidents will also result in a registration problem.

Using definition 1.1, and by chosing one of the images to be the reference R, and the other image to be the deformable template T, one can write that image registration is finding a transformation $\varphi : \mathbb{R}^d \to \mathbb{R}^d$, such that the reference image R and the deformed template image T_φ are similar:

$$T_\varphi(x) := T \circ \varphi(x) = T(\varphi(x)). \qquad (5.1)$$

Eq. (5.1) illustrates the case where R and T are images of the same object. In many applications, the reference and template images show different but similar objects. For example, the reference R can be the X-Ray image of the right hand of an adult male subject, while the deformable template T is the hand of another subject. In this case a measure for images similarity must be defined to solve the image registration problem. For example, a distance measure $\mathcal{D}$ which is a function of image intensities could be chosen as a similarity measure. The image registration problem can then be formulated as follows:

Definition 5.2. Given a distance measure $\mathcal{D} : \text{Images}(d)^2 \to \mathbb{R}$ and two images $R, T \in \text{Images}(d)$, image registration is finding a mapping $\varphi : \mathbb{R} \to \mathbb{R}$ that minimizes $\mathcal{D}(R, T \circ \varphi)$ [12].

Nevertheless, there might be no correspondence between intensities $R(x)$ and $T_\varphi(x)$ for an optimal φ. That can be the case if R and T were taken using different imagine devices for example. A mapping $g : \mathbb{R} \to \mathbb{R}$ can be used on $T_\varphi(x)$ allowing the comparison of R and $g \circ T_\varphi(x)$. Furthermore, distance measures not related to intensities but rather to mutual information of images can also be considered.

With that in mind, the general registration problem can be re-defined as:

Definition 5.3. Given a distance measure $\mathcal{D} :$ Images$(d)^2 \to \mathbb{R}$ and two images $R, T \in$ Images(d), image registration is finding a mapping $\varphi : \mathbb{R} \to \mathbb{R}$, and a mapping $g : \mathbb{R} \to \mathbb{R}$ that minimizes $\mathcal{D}(R, g \circ T \circ \varphi) = \mathcal{D}(R, g(T_\varphi(x)))$ [12].

Although the problem of image registration is easy to state, it is hard to solve. Mathematically speaking, the problem is ill-posed. Therefore, a direct solution approach is impossible.

A general approach to image registration is based on a similarity measure assessment, and regularization. The similarity measure can be seen as the force driving the registration process whereas the regularization as the counter-force that controls and guides the transformation[12].

Since image registration is an ill-posed problem, regularization is essential and inevitable. The regularizer is used to pick the most likely registration. It can also be used to supply additional knowledge in order to guide the registration process.

5.2.2 Image Registration Methods

Unfortunately, no unified treatment or general theory for image registration has been established yet. Different methods and techniques for image registration, each focusing on certain aspects of the application area where the method is being used has been developed separately. These aspects can differ dramatically.

For example, in real time applications used in industrial tracking or inspection, computation time is of great consideration. In applications where high resolution 3D images processing is needed the method must focus on the memory available for the computation. In medical application like medical treatment planning accuracy is a extremely crucial element of the image registration method [301].

The various kinds of information, features and data patterns in images, originating from the different fields where image registration is used, are additional sources of registration methods variations.

Despite all the variability of the image registration methods, the definition of image registration (Def. 5.1) remains valid for all these methods. The kind of *transfor-*

mation as well as the notions of *optimal* and *corresponding* differ depending on the application.

As mentioned above, image registration can be spilt in two parts, a similarity measure and a regularizer. Using this general approach, it is possible to classify the different registration methods in two main categories: parametric image registration and non-parametric image registration [12, 301].

5.2.2.1 Parametric Image Registration

In all methods falling into this category, regularization is achieved by restricting the transformation φ (see Def. 5.3) to a parametrizable set of basis functions ϕ_j with $(j = 1 \dots n)$. While the similarity measures are represented by different distance measures, the transformation parameters α_j are obtained by minimizing a distance measure in the space spanned by the basis functions ϕ_j using algebraic equations or appropriate optimization techniques. Minimizing a distance measure over a parameterized spline space is one of if not the most commonly used registration technique.

Landmark-based, principal axes-based and optimal linear registration methods belong to this category.

In landmark-based registration methods, a number of points i.e. landmarks in the reference image and their corresponding points in the template image are defined. The transformation is obtained by minimizing a distance measure, like the Euclidean distance, between landmarks in the reference and deformed template image. These landmarks could be anatomical landmarks, fiducial markers, or any other features the image contains. Regardless of what these landmarks signify, the computed transformation relates these features. The registration is solely based on these features, thus no further knowledge is taken into account.

The Iterative Closest Point (ICP) method introduced by [302, 303] can be affiliated to this group. The ICP algorithm and the various aspect of this method will be detailed in 5.2.3.

Principal axes-based methods are similar to landmark-based methods, they differ in that geometrical features, like the center of gravity, the standard deviations, and the principal axes of the image objects represented for example in density classes, are used for the registration instead of landmarks. These features can be deduced automatically and thus no user interaction is needed. That provides a main advantage over landmark-based methods. Nevertheless it is not suitable for multimodal densities or multimodal images.

In optimal linear registration methods, distance measures based directly on image intensities are used. Distance measures like the sum of squared differences, correlation-based distance measures, and mutual information make part of this sub-category of parametric registration methods.

B. Zitova *et al.* presented a literature survey of image registration methods [304], most of which belong to the parametric registration methods category. The survey is comprehensive and provides a tremendous starting point for further readings.

5.2.2.2 Non-Parametric Image Registration

As mentioned above, to solve the ill-posed registration problem, regularization is needed. In methods belonging to this category, registration is no longer restricted to a parametrizable set of basis functions. Instead, a regularization term or a regularizer S is used to circumvent the ill-posedness and pick the transformation which is most likely.

Regularizers can also be used to supply additional information or knowledge beside the image information, to further control the transformation in order to obtain a better registration.

All methods belonging to this category can be explained based on the regularized minimization of a particular distance measure. And they are all distinguished by the regularizers they use.

Methods like the elastic registration, fluid registration, diffusion registration and the curvature registration methods make parts of this category.

In elastic registration, images are interpreted as elastic objects. The distance measure provides the force that drives the deformation, while a regularizer S, based on a model of elasticity, restricts the transformation. S is a function of the elastic objects' points coordinates, as well as the distance measure D. The deformation is obtained by minimization of a cost function E which is a sum of a distance measure term E_D and a regularization term E_S for the coordinates of the elastic objects. This method will be detailed in 5.2.4.

The mathematical setting of fluid registration methods is similar to that of elastic registration methods. The difference is that a model of fluid mechanics is used instead of a model of elasticity and the regularizer S is a function of the objects coordinates velocities. Using this method, it is possible to deform any template image to any reference image. Although this feature is helpful for some applications, it is not for others, like when applying fluid registration on an elastic object.

A mathematical formulation and general solution schemes for all mentioned registration methods can be found in [12].

5.2.3 The Iterative Closest Point (ICP) algorithm

The Iterative Closest Point (ICP) algorithm is an iterative process, where in each iteration k two main tasks are performed: The first, is the definition of landmarks in a template T and a reference R image. The second is the registration of the images T and R using the landmarks defined in that iteration k.

5.2.3.1 The Original ICP

The ICP algorithm was first introduced by Chen and Medioni [302] and Besl and McKay [303] to estimate the rigid transformation of roughly aligned 3D data sets. This algorithm is called the original ICP to distinguish it from all the ICP variants that were developed later.

In the original ICP, starting with two 3D point clouds, a template cloud T_c and a reference cloud R_c, landmarks are set in each iteration k by defining correspondence pairs between points of T_c and R_c.

Each correspondence pair consists of a point $\mathbf{x}_{i,k}$ $(i = 1, \ldots, m)$ from T_c and a point $\mathbf{y}_{j,k}$ $(j = 1, \ldots, n)$ from R_c that satisfies the correspondence metric minimal, which is the Euclidean distance $d_{i,k}$ in the original ICP case:

$$d_{i,k} = \min_{j \in \{1,\ldots,n\}} \|\mathbf{y}_{j,k} - \mathbf{x}_{i,k}\| \tag{5.2}$$

After the definition of correspondence pairs, the algorithm searches for the rigid transformation φ_k that minimizes the mean square error of the estimated correspondence pairs e_k:

$$e_k = \frac{1}{m} \sum_{i=1}^{m} \|\mathbf{p}_{i,k} - (\mathbf{R}_k \mathbf{x}_{i,k} + \mathbf{t}_k)\|^2 \tag{5.3}$$

Where $\mathbf{p}_{i,k}$ is the point $\mathbf{y}_{j,k}$ from R_c corresponding to the point $\mathbf{x}_{i,k}$ from T_c, and where $\mathbf{R}_k$ and $\mathbf{t}_k$ are the rotation and the translation parts of φ_k and are given by

$$\mathbf{R}_k = \begin{bmatrix} r_{11} & r_{12} & r_{13} \\ r_{21} & r_{22} & r_{23} \\ r_{31} & r_{32} & r_{33} \end{bmatrix} \text{ and } \mathbf{t}_k = \begin{bmatrix} t_{11} \\ t_{21} \\ t_{31} \end{bmatrix} \tag{5.4}$$

To minimize e_k and thus determine φ_k, Besl and MacKay used the quaternions solution method [305]. Other solution methods like the orthogonal matrices method [306] or the singular value decomposition method [307] could also be used. Using the resulting transformation φ_k, the coordinates of the points $x_{i,k}$ are updated and prepared for the next iteration $k + 1$ according to

$$x_{i,k+1} = \mathbf{R}_k \cdot x_{i,k} + \mathbf{t}_k \tag{5.5}$$

The algorithm iterates until the mean square error e_k falls under a predefined threshold ζ or a maximum number of iterations $k_{\max}$ is reached.

5.2.3.2 ICP Algorithm Variants

Many variants of the ICP algorithm that tackle all aspects of the algorithm have been proposed. Based on a literature survey [308], the different variants of the ICP algorithm can be classified according to six different criterions:

1. Selection of points in template or reference points cloud, or in both clouds for correspondence definition
2. Matching or finding correspondence pairs
3. Weighting the correspondence pairs appropriately
4. Rejecting certain pairs
5. Choosing a distance measure based on the correspondence pairs
6. Minimizing the distance measure

Selection of Points

In the original ICP, all available images points were used for the correspondence pairs definition [303]. A Uniform subsampling of the available images points was used in [309], while random sampling with a different sample of points at each iteration was used in [310]. In variants that use per-sample color or intensity to aid the registration, the selection of points with high intensity gradient was introduced in [311].

Motivated by the observation that small features in the reference are vital for a correct alignment, Rusinkiewicz *et al.* proposed a sampling strategy called Normal-Space sampling that choses points such that the distribution of normals among selected points is as large as possible [308]. Normal-space sampling selects a large number of samples in the area of these features in comparison with uniform and random subsampling, that leads to a better alignment specially in the presence of noise and distortion in the template cloud.

Finding Correspondence Pairs

To define correspondence pairs, the minimal Euclidean distance was used in [303] as a correspondence metric (Eq. 5.2). D. Simon used the same correspondence metric along with a k-d tree to speed up the search process [312]. In the work of Chen and Medioni, the minimal Euclidean distance between the template point and the intersection of the ray originating at the template point in the direction of the template point's normal with the reference surface was used [302]. Projecting the source points on the destination mesh from the point of view of the destination mesh's range camera was used in [313, 314]. Another method is to project the source point onto the destination mesh, then to perform a search in the destination range image like in [311, 315]. It is also possible to restrict the search for correspondence pairs only to points from the reference cloud compatible with the temaplate points, the compatibility could be based on color as in [316], or angel between normals as presented in [317].

Weighting of Correspondence Pairs

By looking back to Eq. (5.3), the equation can be updated to incorporate weights $w_{i,k}$ associated with the pairs $(\mathbf{x}_{i,k}, \mathbf{p}_{i,k})$:

$$e_k = \frac{1}{m} \sum_{i=1}^{m} w_{i,k} \cdot \| \mathbf{p}_{i,k} - (\mathbf{R}_k \mathbf{x}_{i,k} + \mathbf{t}_k) \|^2 \tag{5.6}$$

By comparing both Eqs. (5.3) and (5.6) we can say that the original ICP uses constant weights $w_{i,k} = 1$. Weights based on the compatibility of normal vectors (the inner product the normals), or on the compatibility of colors [316] have been used. Assigning lower weights to pairs with greater point to point distances has been introduced in [316].

Rejecting Certain Correspondence Pairs

Variants of ICP apply the rejection of correspondence pairs to eliminate outliers. Different strategies were proposed like rejecting pairs with a distance larger than a given threshold or larger than a multiple of the standard deviation of distances, or rejecting a percentage of pairs based on a specific metric. Also pairs that are not consistent with neighboring pairs [318] can be rejected. Turk *et al.* proposed the rejection of pairs that contain mesh boundaries [309].

Chosing Distance Measures

In the original ICP the minimization of the squared Euclidean distance between corresponding points is used as a distance measure (Eq. 5.3). Other distance measures use the sum of squared distances from each source point to the plane containing the destination point and oriented perpendicular to the destination normal [302]. In the trimmed ICP variant [319] squared Euclidean distances are sorted and distances bigger than a certain value are not included.

Minimizing the Distance Measure

Solution methods for minimization of the squared Euclidean distances like the singular value decomposition [307], quanternions[305], orthonormal matrices [306], and dual quaternions [320] have been used. An attempt to solve the problem using the Levenberg-Marquardt algorithm was presented in [321]. In the trimmed ICP variant, the minimization of a sum of the sorted squared Euclidean distances that are smaller than a specific value was compared with minimizing the median of the sorted squared Euclidean distances method [322].

5.2.3.3 ICP Computational Performance

For the original ICP, the complexity of a single iteration is $O(m \cdot log(n))$ with m the number points of the template cloud T_c and n the number points of the reference cloud R_c. Because the ICP algorithm finds the minimum distance between the two points clouds iteratively, the overall complexity is $O(k \cdot m \cdot log(n))$, where k is the iterations count [303]. Therefore the registration of high-resolution images of both template and reference images with the original ICP algorithm is a heavy computational task.

There are two main approaches to speed up the ICP algorithm:
1. Reducing iterations count
2. Reducing correspondence pairs definition time

The first approach is reducing the number of iterations needed for the algorithm to reach a final state. Besl and McKay introduced a method to reduce the number of iterations by updating the motion parameters using linear or parabolic extrapolation [303]. A. Almhdie *et al.* proposed a new implementation of the ICP algorithm by using a complete distance matrix during the process of definition of correspondence pairs [323]. In his work Almhdie states that his implementation provides a faster convergence, in terms of iterations count, a more precise estimation of pair of points correspondence, and a better resilience to additive Gaussian noise and outliers.

The second approach is to reduce the search time for finding the correspondence pairs, a problem known as the nearest neighbor problem. Finding correspondence pairs is the most time consuming step of the ICP algorithm. Therefore, improving the computation time of finding the nearest neighbor was the focus of many works.

The nearest neighbor problem is a classical problem that occurs in many applications. This problem has general solutions, like the k-d tree data structure [324], "Elias" [325], as well as other methods [326, 327, 328]. One of these solutions is usually implemented with ICP, however, the k-d tree data structure suggested by [303] is the most widely used [312, 317].

Algorithms, specially developped to be used with ICP implementations, like the Spherical Triangle Constraint Nearest Neighbor (STCNN) [329], and the Hierarchical Model Point Selection/Logarithmic Data Point Search (HMPS/LDPS) [330], and methods that use application-specific image properties [331], were also suggested.

5.2.3.4 Nonrigid ICP

The ICP algorithm is widely used to perform rigid image registration, due to its simplicity and robustness. For the same reasons, and the big interest in nonrigid image registration, attempts has been made to extend the ICP algorithm to support nonrigid registration.

In the early days of ICP, Feldmar and Ayache developed a method for nonrigid surface registration but introducing various extensions to the original ICP algorithm [332]. In their work, the nonrigid registration was conducted in stages, first a rigid displacement is calculated, then the best affine transformation is obtained, at the end the surface is deformed by transforming each point of the surface with a local affine transformation which is varying smoothly along the surface. Obviously, the last stage is the stage responsible for the nonrigid part of the registration. As mentioned above, an affine transformation must be calculated for each point. Therefore, for each source point a spherical search region is defined, the closest target point is associated with the current investigated source point, and the transformation is the one that minimizes the square distance between the point pairs. after all transformations are calculated, the rigid displacement of each point is smoothed using weighted displacements of other points in the spherical search regions of that point. The weights decrease linearly with the distance from the sphere center.

Since then, several other methods have been developed [333, 334, 335]. Amberg *et al.* [335] showed how to extend the ICP algorithm to nonrigid registration of 3D surfaces while retaining the convergence properties of the original algorithm. Regularization based on minimizing the difference between transformations acting on neighboring mesh vertices was used in that work. This regulization is very similar to the regularization used in elastic registration. Indeed, the term stiffness was frequently used in [335], making this algorithm a non-parametric registration method.

As in all ICP algorithms, the registration is made by calculating successive transformations. Each transformation was calculated by minimizing a cost function E that is the sum of several terms:

$$E(\mathbf{X}_k, k) = E_d(\mathbf{X}_k) + E_\lambda(\mathbf{X}_k) + \alpha_k \cdot E_s(\mathbf{X}_k) \tag{5.7}$$

where k is the current iteration, $\mathbf{X}_k$ is the template points coordinates in iteration k and α_k is a stiffness weight that decreases with increasing k ($\alpha_k > \alpha_{k+1}$). E_d is the correspondence pairs distances cost function term and can be given with

$$E_d(\mathbf{X}_k) = \sum_{i=1}^{m} d_{k,i}^2 \tag{5.8}$$

where $d_{k,i}$ can be calculated using Eq. (5.2). Similarly, the landmarks cost function term E_λ can be given by

$$E_\lambda(\mathbf{X}_k) = \sum_{j=1}^{l} d_{k,j}^2 \tag{5.9}$$

where l is the number of the landmarks, $d_{k,j} = \|\mathbf{p}_j - \mathbf{x}_{k,j}\|$, $\mathbf{p}_j$ is the landmark j in the reference image, and $\mathbf{x}_{k,j}$ is the corresponding landmark in the template image in iteration k. Finally, E_s is a stiffness term that regularizes the deformation by penalizing the weighted difference of the transformations of neighboring vertices.

The algorithm iterates over k and thus over a set of decreasing stiffness weights, and incrementally deforms the template towards the reference, recovering the whole range of global and local deformations.

5.2.4 Elastic Image Registration

Elastic image registration makes part of the non-parameteric registration category (see Section 5.2.2.2), where objects displayed in images are interpreted as elastic objects. The differences between objects in the template and corresponding objects in the reference drive the registration while a model of elasticity, like lineary elastisity, regularizes the deformation. For a good mathematical formulation of regularization in elastic image registration, the reader is directed to the work of Modersitzki [12].

Elastic registration is suitable for registration tasks where the elastic properties of the objects depicted in the images should be taken into account.

Elastic image registration has already been used for a wide range of applications. Allen *et al.* [336] used elastic image registration to fit high-resolution template meshes to detailed human body range scans with sparse 3D markers. The Nonrigid ICP algorithm presented by Amberg *et al.* [335] (see Section 5.2.3.4) uses the same cost function used in [336] and thus it is also an elastic registration method. N. Hasler *et al.* [337] presented a unified model that describes human pose and body shape using an elastic image registration method very similar to that in [336, 335] with some differences in the distance measure definition.

The Finite Difference Method (FDM) and the Finite Elements Method (FEM) are widely used to solve the equations arising from the mathematical models of elasticity [338, 12, 339, 301, 340].

5.3 Elastic Image Registration with Morphi

Similar to the nonrigid ICP algorithm of Amberg *et al.* [335] described in Section 5.2.3.4, an elastic image registration method, that uses the modified mass-spring system (Adamss) (see Chapter 2) to model elasticity, was developed in this work. The framework, comprising the implementation of this method, was named Morphi.

Unlike other elastic registration methods, where models of isotropic linear elasticity are employed, the elasticity model used in this work (Adamss) is able to simulate non-linear, anisotropic behavior. Furthermore, it allows for volume preservation or volume control during the image registration process. By remembering that the model of elasticity serves as a regularizer in the registration process, one can see that these features increase the control over the registration process and thus improve its quality.

Since Morphi is based on the ICP algorithm, it is an iterative process, where in each iteration k correspondence pairs are defined from the template image T and reference image R, then a mapping φ_k between template and the reference images is calculated. And this procedure iterates until a stop criterion is reached.

As mentioned in Section 5.2.3.2, ICP algorithms differ in several aspects.

In this work, points from the template T and the reference R were selected prior to the registration process, to form the template and the reference points clouds, T_c and R_c respectively. This is needed for the definition of correspondence pairs. In the general case, T_c and R_c have different number of points.

T_c is divided to a number of subsets of points. Accordingly, R_c is divided to subsets that correspond to the template cloud subsets. By assigning features in T and R, important for the registration, to different subsets of T_c and R_c, the quality of the registration can be improved by allowing correspondence pairs to be built only between points belonging to the same subsets in T_c and R_c.

To define the correspondence pairs between a subset of T_c and a subset of R_c the minimal Euclidean distance, given in Eq. (5.2), was used as a correspondence metric.

Different weights w_s were associated with each subset s of the points clouds, and as a distance measure the weighted sums of squared Euclidean distances between corresponding points was used as given in Eq. (5.11).

Corresponding landmarks could also be make part of T_c and R_c.

To calculate the mapping φ_k, a cost function $E(\mathbf{X}_k, k)$ simular to that of Allen *et al.* [336] and Amberg *et al.* [335] must be minimized in each iteration k. The cost function used in this work is given by

$$\mathbf{E}(\mathbf{X}_k, k) = \mathbf{E}_\epsilon(\mathbf{X}_k) + \mathbf{E}_{icp}(\mathbf{X}_k, k) + \mathbf{E}_\lambda(\mathbf{X}_k, k) \tag{5.10}$$

where $\mathbf{X}_k$ are the vertices coordinates of the template at iteration k. $\mathbf{E}_\epsilon(\mathbf{X}_k)$ is related to the model of elasticity (Adamss) and can be interpreted as the set of strain energy density functions of the model governing the different passive elastic properties of the model, that is in turn reflected in the model's stiffness. $\mathbf{E}_{icp}(\mathbf{X}_k, k)$ is the ICP algorithm distance measure and is given by

$$\mathbf{E}_{icp}(\mathbf{X}_k, k) = k \sum_{i=1}^{m} w_{s,i} d_{k,i}^2 \tag{5.11}$$

where m is the count of the source points, $w_{s,i}$ is the weight specific for the subset to which $\mathbf{x}_i$ belongs; $d_{k,i}$ is the distance between the corresponding pair $(\mathbf{x}_i, \mathbf{x}_j)$ as given in Eq. (5.2).

Similary, the landmarks term $\mathbf{E}_\lambda(\mathbf{X}_k, k)$ is given by

$$\mathbf{E}_\lambda(\mathbf{X}_k, k) = k \sum_{l=1}^{L} w_l d_{k,l}^2 \tag{5.12}$$

where L is the count of the landmarks, w_l is a weight specific for each landmark and $d_{k,l}$ is the Euclidean distance between the template landmark and its corresponding reference landmark at iteration k.

The cost function $\mathbf{E}(\mathbf{X}_k, k)$ can be minimized by setting

$$\frac{\partial \mathbf{E}(\mathbf{X}_k, k)}{\partial \mathbf{X}_k} = 0 \tag{5.13}$$

and solving for $\mathbf{X}_k$. An implementation of the Newton's method for solving non-linear systems in the Portable, Extensible Toolkit for Scientific Computation (PETSc) [59, 341] was used for solving the resulting non-linear system.

The resulting φ_k is used to transform the template image to a new configuration in iteration $k + 1$ and the procedure iterates until a stop criterion is reached. The stop criterion could be set to a predefined maximum iterations count. Alternatively, the iteration could be stopped, when the mean square error of distances between corresponding pairs, including landmarks, reached a predefined minimum threshold.

5.4 Simulations

In the following sections, three different applications using Morphi for image registration are presented. Details about the specific implementation and the results are presented for each of the applications.

5.4.1 Heart Modeling Validation with Morphi

The use of the elastic image registration method for the validation of heart modeling simulations was mentioned in Section 4.7.7.2. In the following the procedure used for the validation is detailed.

First, a simulation of ventricular deformation is conducted as described in Chapter 4.

The resulting deformation at systole is then used as the template image T in the elastic registration procedure, whereas another segmented dataset of the same ventricles, also in systole state, is used as the reference image R.

The template points cloud T_c is generated by selecting points from the endocardial and epicardial surfaces of both ventricles in T while the reference points cloud R_c is generated by selecting points from the endocardial and epicardial surfaces of both ventricles in R.

To improve the quality of the registration, the points of T_c and R_c are grouped in four subsets: The endocardial and epicardial surface of the left and right ventricle.

Points belonging to one of the template cloud subsets will search for correspondences only in the respective reference cloud subset.

Landmarks in T and R are also set to obtain a better registration and thus a better validation.

The iterative registration procedure iterates until the mean distance $\bar{d}$ of all correspondence pairs and landmarks is below a specific threshold $\bar{d}_{\mathrm{th}}$, and the maximum distance d_{max} is below another threshold $d_{\mathrm{max,th}}$.

After the registration task is completed, the displacement of the elastic model nodes can be used to describe the similarity between the ventricular deformation simulation in systole state and the segmented dataset also at systole. Statistical methods can be applied to these displacements to generate simplified measures or indicators that describe the similarity.

In this work, this method was used to validate simulations: 1, 4-a, 4-b and 4-d (see Table 4.4), detailed in Section 4. The templates were generated from the simulation $260ms$ after the beginning of the systole and the reference was segmented from the Cine sequence dataset (see Section 4.7.1.2) at $280.8ms$ after the R-Peak.

A second validation of simulation 4-d was repeated using 23 tagging landmarks extracted from FLASH sequence (see Section 4.7.1.2).

In all simulations with Morphi described here, the following settings and parameters were used for the model of elasticity: Both ventricles in the templates were considered to be nearly incompressible Neo-Hookean material (see Eq. 3.4), with $c_1 = 2 \times 10^2\ kPa$, $\beta = 1 \times 10^2\ kPa$ and density $\rho = 100\ kg/m^3$.

All subset weights w_{s} in Eq. (5.11) were set to the same value $w_{\mathrm{s}} = 10$. Foa the simulations that included landmarks, the landmarks weights w_{l} were all set to the same value $w_{\mathrm{l}} = 10 w_{\mathrm{s}}$ to emphasize their influence on the registration outcome. Finally, the values $\bar{d}_{\mathrm{th}}$ and $d_{\mathrm{max,th}}$ were set to $0.07\ mm$ and $0.5\ mm$, respectively. After the registration processes were done, the nodes displacements were used to compare the ventricular deformation simulations 1, 4-a, 4-b and 4-d with and without landmarks, by calculating the Euclidean norms of all displacements, the mean $\bar{d}$ and standard deviation STD of the Euclidean norms.

Table 5.1 shows the calculated mean $\bar{d}$ and standard deviation STD of the displacements of the right and left ventricles of all conducted validation registration tasks. Because no detailed information on the movement of the ventricles were available, specially about the rotation of the ventricles, the registration results represent the similarity between the shape of the modeled ventricles at $t = 260ms$ and the ventricles segmented from the Cine sequence at systole regardless of rotation.

Table 5.1 Registration tasks conducted with Morphi for the validation of heart modeling. The mean $\bar{d}$ and standard deviation STD of the displacements of the particles are listed here. The values were calculated for the right ventricle, the left ventricle and for both right and left ventricles.

Registration ID	T from Simulation ID	Landmarks	$d \pm STD(\text{mm})$
1	1	-	2.12308 ± 1.66308
2	4-a	-	6.25605 ± 2.93281
3	4-b	-	2.63679 ± 2.16585
4	4-d	-	4.31644 ± 2.87201
5	4-d	23	4.27799 ± 2.86351

Registration ID: 1 showed the lowest mean distances followed by registration ID 3, 5, 4 and finally 2 which came last among the registration tasks. The table shows that using the 23 landmarks in the left ventricle in registration ID: 5 provided a slight improvement in the quality of registration over the results of registration ID: 4.

Figures 5.2 show the ventricles resulting from the different deformation simulations at $t = 260ms$ colored according to the Euclidean norms of the displacements of particles. Blue indicates a low Euclidean norm while red indicates a high Euclidean norm. Figures 5.3 demonstrates the effect of using landmarks on the registration.

5.4.2 Tracking Ventricular Deformation with Morphi

Computer models of electrophysiology are used to estimate the body surface electrocardiogram (ECG). Anisotropic heart and torso models used to solve the forward problem of electrocardiography do not account usually for the heart deformation during a heart cycle. To investigate and quantify the effects of ventricular motion on the body surface electrocardiogram (ECG), especially during the ST segment and the T-Wave, the ventricular motion must be taken into account during forward calculation. This investigation was made in collaboration with David Keller and Thomas Fritz at the institute of biomedical engineering (IBT) at the Karlsruhe institute of technology (KIT) [342].

Since the forward calculation requires the definition of a mesh and cardiac potential sources for the entire simulation, each cardiac potential source must be tracked, that means the trajectories of each cardiac potential source must be known at each timestep of the cardiac cycle.

To obtain these trajectories that describe the displacement of the nodes in the unstructured tetrahedron mesh used for the forward calculation, a deformation field

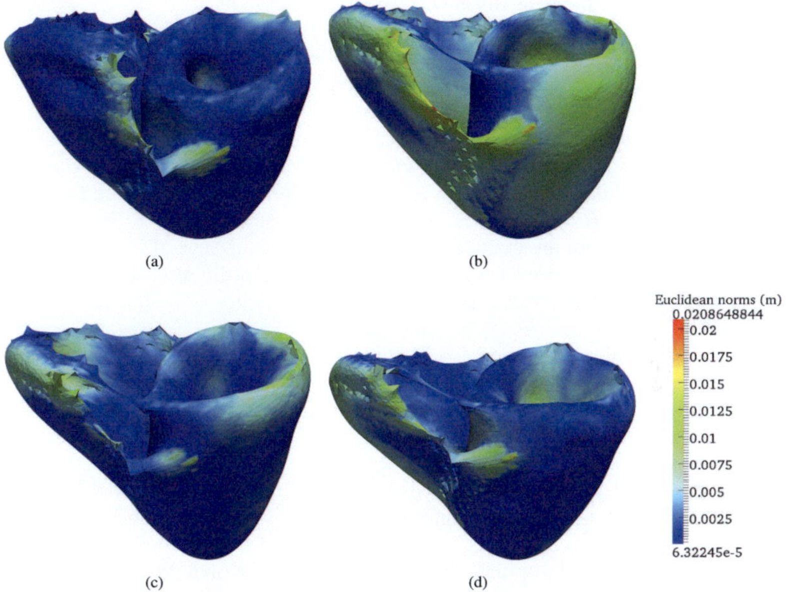

(a) (b)

(c) (d)

Fig. 5.2 The distribution of Euclidean norms of the displacement of particles of registrations 1 (a), 2 (b), 3 (c) 4 (d) are depicted in the figures. The ventricles resulting from the deformation simulations at $t = 260ms$ are colored according to the Euclidean norms of the displacements of particles. Registration 1 shows the best results while 2 shows the worse results.

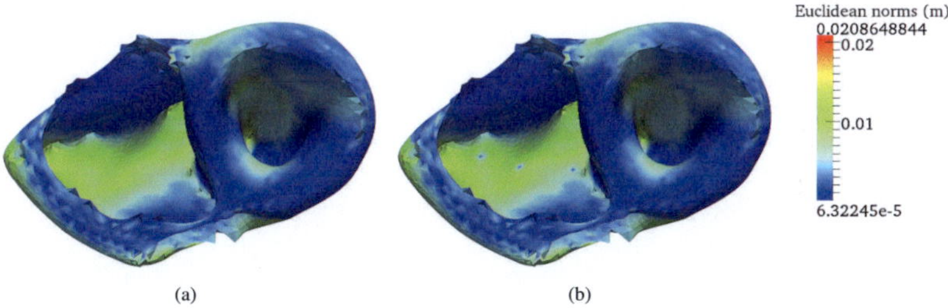

(a) (b)

Fig. 5.3 A top view on the ventricles in registration task 4 (a) and 5 (b), the effect of 3 of the landmarks used in registration 5 can be seen clearly on the wall of the right ventricle where 3 blue dots can be seen. 23 landmarks were used in this registration task that slightly improved the registration result.

is needed for each of the time instants.

In this work, Morphi was used to obtain these deformation fields based on MRI data of a volunteer. The used method is described in the following.

A segmented dataset of the heart and the surrounding organs should be used as the template image T in the elastic registration procedure. Several segmented datasets of the same heart and the surrounding organs in n contracted states should be used as reference images R_i $(i = 1, \ldots, n)$. The template points cloud T_c is generated by selecting points from the endocardial and epicardial surfaces of both ventricles in T. n reference points clouds $R_{c,i}$ are generated by selecting points from the endocardial and epicardial surfaces of both ventricles in R_i.

Similar to the method in Section 5.4.1, points of T_c and $R_{c,i}$ are grouped in four subsets: the endocardial and epicardial surface of the left and right ventricles to improve the quality of the registration. Additionally, landmarks are set in T and R_i to obtain a better registration and thus better validation.

By using the different references R_i, n registration tasks are performed and thus n deformation fields at the corresponding contracted states are generated. Deformation fields, in contracted states other than the n states used in generating the various R_i, are linearly interpolated between the deformation fields resulting from elastic image registration.

In this work the segmented diastolic dataset containing the heart and surrounding organs described in Section 4.7.1.2 was used as a template T, and 29 datasets of the same heart segmented from the Cine sequence also described in Section 4.7.1.2 were used for R_i $(i = 2, \ldots, 30)$ where i is the sequence index in the Cine dataset.

It was assumed that the movement of the ventricles only affects tissues in their immediate vicinity. With this assumption in mind, a bounding box of $180 \times 148 \times 136 \ mm^3$ that contained the heart in diastolic state and surrounding tissues (fat, muscle, lung, liver, spleen, colon) was introduced. The movement of all tissues within this box was considered during elastic image registration procedures.

For all 29 image registration procedures described in this section, the following setting and parameters were used with the model of elasticity.

A nearly incompressible Neo-Hookean material model was used to describe myocardial tissue (see Eq. 3.4), with $c_1 = 2 \times 10^2 \ kPa$, $\beta = 1 \times 10^2 \ kPa$ and density $\rho = 100 \ kg/m^3$. The same model was used for all other materials as well but with $c_1 = 2 \times 10^2 \ kPa$, $\beta = 1 \times 10^{-4} \ kPa$ and density $\rho = 100 \ kg/m^3$.

All subset weights w_s in Eq. (5.11) were set to the same value $w_s = 10$. The weights of landmarks w_l were set to the same value $w_l = 10 \times w_s$ to emphasize their influence on the registration outcome.

Finally, the values $\bar{d}_{th}$ and $d_{max,th}$ were set to $0.07\ mm$ and $0.5\ mm$, respectively.

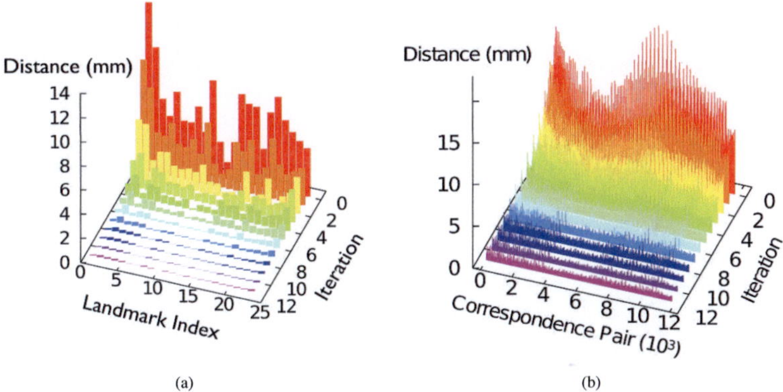

Fig. 5.4 Euclidean distances of the landmarks from T_c to $R_{c,14}$ (a), and of all points from T_c to their closest points in $R_{c,14}$ (b). The distances decrease with each iteration. After 11 iterations, the distances of all points were below the stop criterion.

In all registration procedures, the stop criterion was reached after 8 to 12 iterations.

Figure 5.4(a) shows how the Euclidean distances $d_{k,l}$ between the template landmarks and their corresponding reference landmarks changed during one of the registration procedure while Figure 5.4(b) shows the Euclidean distances $d_{k,i}$ between all points i of T_c including landmarks and their corresponding counterparts from $R_{c,14}$ for all iterations k.

Figures 5.5 show the mean and the standard deviation of the Euclidean distances before and after the registration for all conducted registration procedures of landmarks (Fig. 5.5(a), and of all points in T_c including landmarks (Fig. 5.5(b)).

To quantify rotation of the landmarks during ventricular contraction, their rotation around the longitudinal axis of the left ventricle was calculated. The average systolic rotation was $4.7° \pm 2.5°$ with a maximal and minimal rotation of $10.7°$ and $0.6°$ respectively, which is in the range of reported values in literature [343] at the respective ventricular locations.

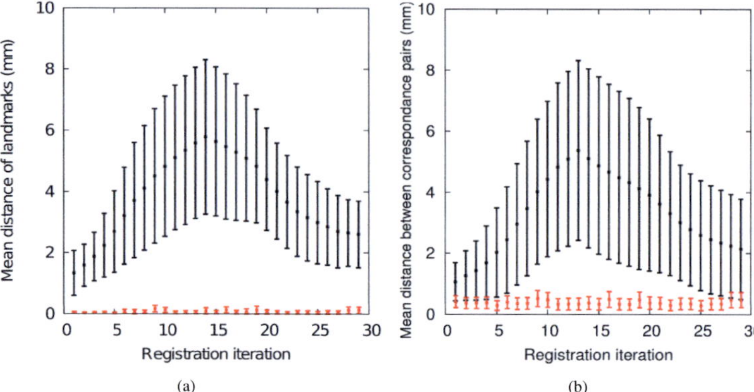

(a)   (b)

Fig. 5.5 The mean distances $\bar{d}$ and the standard deviation (STD) of all 29 registration procedures are shown in the figures before the registration (black) and after the registration (red). $\bar{d}$ before and after the registration tasks were calculated once for the Landmarks (a) and once for all points in T_{c}.

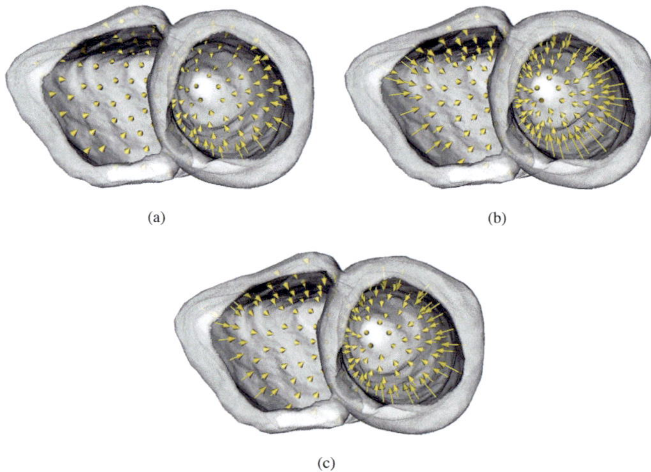

(a)   (b)

(c)

Fig. 5.6 The displacement vectors are visualized together with the transparent diastolic ventricular state in contraction phase at $t = 160ms$ (a), in systole at $t = 311ms$ (b) and in the relaxation phase at $t = 462ms$ (c).

5.4.3 Generation of Personalized Torso Models for Detector Efficiency Calibration in Virtual Radiation Protection Scenarios

In-vivo measurements of incorporated radioactive isotopes are performed periodically on workers who encounter open radioactive material. Measurements are also performed in the event of a suspected incident, such as ingestion or inhalation of radionuclides. These measurements aim to estimate the absorbed dose of radionuclides by determining the activity of these nuclides within the body. For the measurements, detectors outside the body are used to count a fraction of the photons emitted from radioactive nuclides within the body e.g. phoswich detectors, which are detectors developed to detect low-intensity, low-energy gamma rays, X-rays, as well as alpha and beta particles efficiently in a higher-energy ambient background. If the detector efficiency η is known, the activity A can be calculated by measuring the count rate cps according to

$$cps = \eta A \tag{5.14}$$

where, cps is the photon $counts/s$, η is given is $counts/decay$ and A in $decays/s$. Once the activity is determined, the absorbed dose can be estimated and the medical treatment can be then adjusted accordingly.

To perform the measurements, a detector efficiency calibration is needed. The precision of the determination of the activity in Whole Body Counter (WBC) and Partial Body Counter (PBC) detectors depends mainly on systematic errors of the detector efficiency calibration. The calibration of both WBC and PBC detectors is performed using measurements of physical phantoms that consist of material that model the density and the effective atomic number of biological tissue in human.

Numerical methods using Monte Carlo methods are being investigated for numerical calibration of detectors efficiencies. A virtual world, comprising the subject body, the detectors and the rules that simulat the photons and electrons physics, can be used to generate virtual in-vivo measurement scenarios.

Using personalized torso models is expected to improve determining detectors efficiencies and thus reducing systematic errors. Individual torso models can be generated by segmentation of three-dimensional medical images which is still an expensive and time consuming task making it unsuitable for large scale implementations. Alternatively, personalized torso models can be generated by using image registration to adapt or personalize already segmented reference models.

Affine parametric image registration was used by L. Hegenbart [344] for the generation of personalized torso models. The detector efficiency for in vivo measurements of low energy photon emitters in the thorax depends strongly on chest wall thickness (CWT). CWT can be defined as the distance from the anterior skin surface of the thorax to the pleura (the thin skin surrounding the lung). This distance is not constant throughout the chest and is mainly determined by the body size, age, muscle and fat tissues. Hegenbart concluded that affine image registration cannot reproduce the CWT with big accuracy, whereas elastic image registration is expected to perform better.

In this work, Morphi was used for the generation of personalized torsos for detector efficiency calibration in virtual radiation protection scenarios. The resulting torso models were used in typical lung counting scenarios to asses the quality of the resulting torso models in calibration of detectors efficiencies. This investigation was made in collaboration with Lars Hegenbart and Stephen Pölz at the institute of radiation research (ISF) at the Karlsruhe institute of technology (KIT).

The segmented torso model is used as the template T. The surface of the segmented torso model is extracted and points on the surface are selected to make the template points cloud T_c. T_c is divided to several subsets marking the back, the chest, and two subsets marking the sections at the bottom and around the neck region in the torso models. A surface scan of an individual is used to generate the reference R. By selecting points of R the points cloud R_c is created. R_c is then grouped in several subsets that correspond to those of T_c.

By assigning different elasticity parameters to the different tissues of T, the registration can be controlled to reproduced a better representation of the subject's torso.

The lungs volume ratio r_lungs, and the liver's volume ratio r_liver can be used as additional information to further improve the quality of the registration. The lungs volume ratio r_lungs, is the ratio of the measured individual's lungs volume V_lungs to the lungs volume of the already segmented model $V_\mathrm{T,lungs}$ and can be given with:

$$r_\mathrm{lungs} = \frac{V_\mathrm{lungs}}{V_\mathrm{T,lungs}} \tag{5.15}$$

r_liver is calculated similarly using the measured volume of the liver V_liver and the volume of the liver in the already segmented model $V_\mathrm{T,liver}$.

If additional information about chest wall thickness were available, further points, defining the surface of the lungs or specific areas on that surface for example, can be added to the T_c and R_c in special subsets.

To validate this method and find the best settings for the registration procedure, simulations were made with two different already segmented torso models where the first was used as a template and the surface of the other as the reference as detailed above. After image registration was done, Monte Carlo detector efficiency calibration simulations were conducted using the registered template. The results were compared to Monte Carlo simulations conducted using the second segmented torso model (not just its surface), to determine the success of the registration and thus the success of the method.

Table 5.2 list the simulations conducted in this work for the validation of the method.

In the first set of registration simulations (B000, B001 and B002), the adult male voxels model of the International Committee for Radiation Protection (ICRP) [345] was used as the template T and the surface of the MEETMan voxel model [11] served as the reference R. Both models were cut to the torso. Head, arms, and legs were removed, and air voxel inside the lungs in both models were assigned different values than air voxels outside the body. Both models are visualized in Figure 5.7.

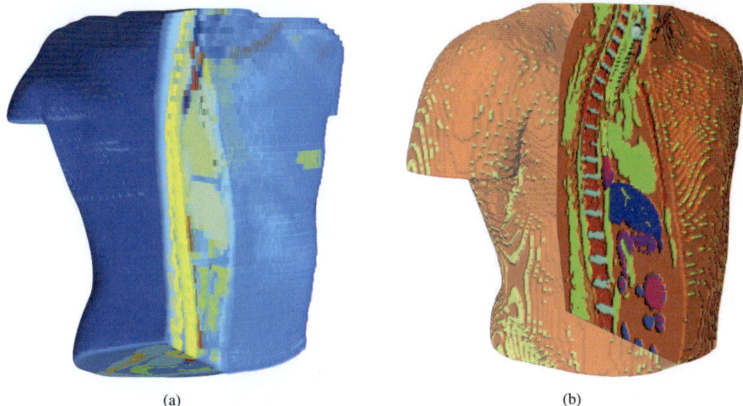

(a)   (b)

Fig. 5.7 The models used for validation of the described method for generation of personalized torso models for detector efficiency calibration in virtual radiation protection scenarios. The surface of the ICRP adult male voxel model (a) was used as the template points cloud T_c, and surface of the MEETMan model (b) was used as the reference points cloud R_c.

In the second set (0054, 0055 and 0056), the surface of the lungs in both the ICRP and the MEETMan models were added to T and R respectively. This was done to accommodate the importance of CWT on the detector efficiency for in vivo measurements of low energy photon emitters. However that requires the segmentation of the individual's lungs which is impractical.

As a simple alternative, the CWT can be estimated using routine ultrasonic measurements on the chest (Fig. 5.8(a)) in the area where the detectors are placed (Fig. 5.8(b)). In the third set of registration simulations (0057, 0058, 0059, 0069, 0070 and 0071), 12 circular areas, 6 for each lung, that represents 12 separate ultrasound measurements of the CWT, were defined as landmarks on the surface of the lungs of the ICRP and the MEETMan models as shown in Figure 5.9 for the ICRP model.

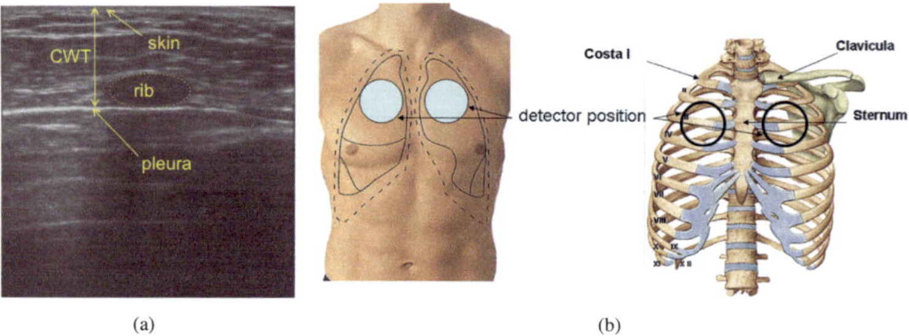

Fig. 5.8 2D ultrasound image (a) of the chest showing the skin used to measure the CWT around the area where the detectors are placed namely between the 3rd and 5th rib on each side of the chest for lung counting scenarios (b) (reproduced from [346]).

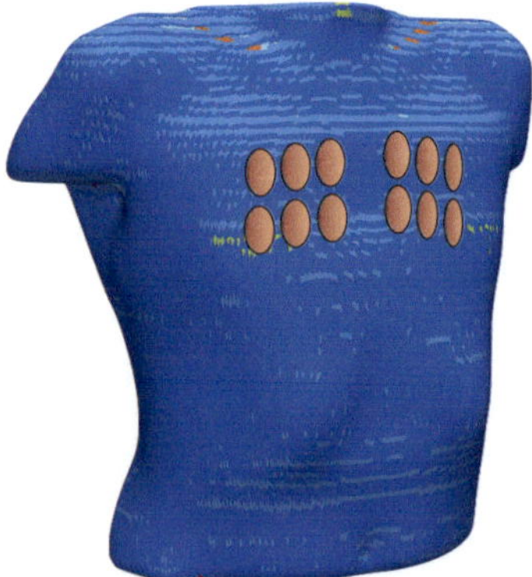

Fig. 5.9 The ICRP adult torso model and 12 circular areas (in orange), 6 on over each lung representing 12 separate ultrasound measurements.

In all sets of simulations described here, the following settings and parameters were used for the model of elasticity (unless otherwise stated) : All tissues were considered to be nearly incompressible Neo-Hookean material (see Eq. 3.4), with $c_1 = 1kPa$, $\beta = 1 \times kPa$ and density $\rho = 1kg/m^3$. All subset weights w_s in Eq. (5.11) were set to the same value $w_s = 10$.

Table 5.2 The registration simulations conducted in this work to validate the method for generating personalized torso models for detectors efficiency calibration.

Reg. ID	T_c (ICRP)	R_c (MEETMan)	Landmarks	Heterogeneity of c_1 (kPa)	Volume Control
B000	Torso surface	Torso surface	-	-	-
B001	Torso surface	Torso surface	-	-	$r_{\text{lungs}} = 1.267$
B002	Torso surface	Torso surface	-	-	$r_{\text{lungs}} = 1.267$ $r_{\text{liver}} = 1.116$
0054	Torso & Lungs surfaces	Torso & Lungs surfaces	-	-	-
0055	Torso & Lungs surfaces	Torso & Lungs surfaces	-	-	$r_{\text{lungs}} = 1.267$
0056	Torso & Lungs surfaces	Torso & Lungs surface	-	-	$r_{\text{lungs}} = 1.267$ $r_{\text{liver}} = 1.116$
0057	Torso surface	Torso surface	12 areas	-	-
0058	Torso surface	Torso surface	12 areas	-	$r_{\text{lungs}} = 1.267$
0059	Torso surface	Torso surface	12 areas	-	$r_{\text{lungs}} = 1.267$ $r_{\text{liver}} = 1.116$
0069	Torso surface	Torso surface	12 areas	$c_{1,\text{lungs}} = 2$	$r_{\text{lungs}} = 1.267$
0070	Torso surface	Torso surface	12 areas	$c_{1,\text{lungs}} = 5$	$r_{\text{lungs}} = 1.267$
0071	Torso surface	Torso surface	12 areas	$c_{1,\text{lungs}} = 5$	$r_{\text{lungs}} = 1.267$ $r_{\text{liver}} = 1.116$

The registration simulations reached the stop criterion within 10-20 iterations. Simulations where additional information like r_{lungs}, r_{liver} or the lungs surfaces were added resulted in better registrations of the lungs and liver. The best overall registration could be observed in registration simulation 0056 where all additional information was delivered to the registration method.

Since the 12 circular landmarks represents a subdomain of the lungs surfaces, the simulations 0057, 0058, 0059, 0069, 0070 and 0071 performed worse than simulations 0054, 0054 and 006 but better than B000, B001 and B002.

An increase of the $c_{1,\text{lungs}}$ in simulation 0069 and a further increase in 0070 and 0071 improved the quality of the third set of registration simulations. Because in simulations 0069, 0070 and 0071 the increased stiffness of the lungs forced more lungs tissue to move with the circular areas landmarks in comparison with simulations 0057, 0058 and 0059.

Figures 5.10 demonstrate the results of registration 0069.

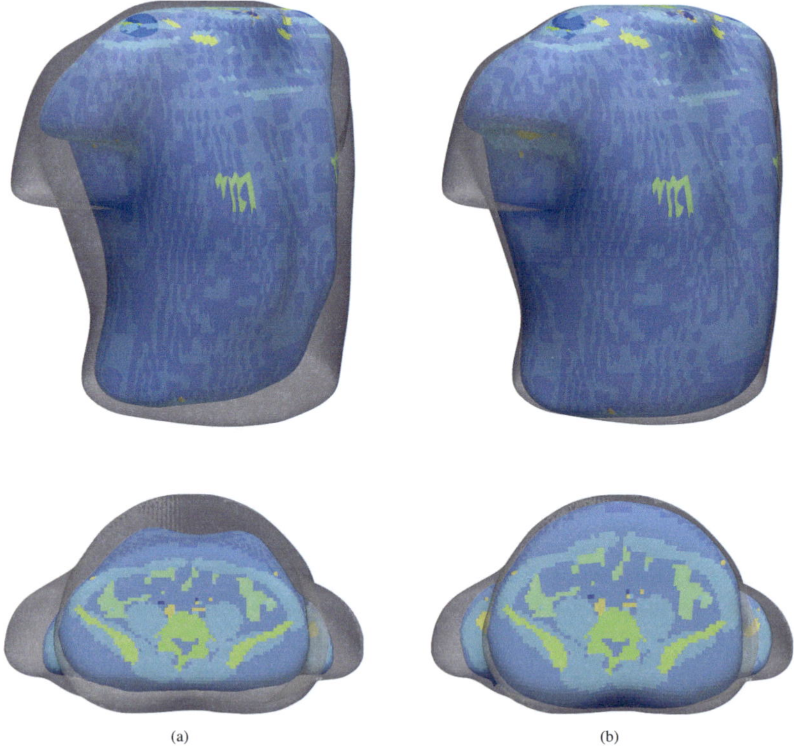

(a)   (b)

Fig. 5.10 The result of registration 0069 are resented here where the ICRP (colored) and the MEETMan (transparent gray) torso models before (a) and after registration (b).

Typical lung counting scenarios with 241Am contamination have been created using the models resulting from the registration. The scenarios included two phoswich detectors placed over centre of mass of the complete lung about 1 mm above skin surface in standard position (i.e. $25°$ inclined horizontally and vertically, approximately parallel to the model surface).

Monte Carlo simulations were conducted to calculate the counting efficiencies as a function of photons energy were recorded. The curves in Figure 5.11 plot the counting efficiencies against the photon energy for the ICRP adult male torso model (red), the MEETMan torso model (green), B001 (blue), 0056 (cyan), 0059 (magenta) and 0069 (brown). The counting efficiency of the ICRP model used as the template T in the registration simulations is about 1.5 times larger than the counting efficiencies of the MEETMan model. The registration resulted in counting efficiencies closer to those of the MEETMan. Using the lungs surfaces for the registration resulted in an efficiency curve (cyan) very close to that of the MEETMan model.

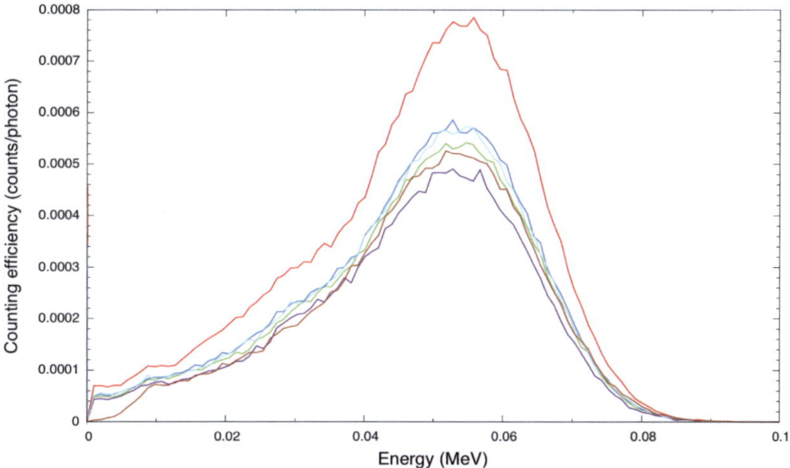

Fig. 5.11 The counting efficiencies for the ICRP adult male torso model (red), the MEETMan torso model (green), B001 (blue), 0056 (cyan), 0059 (magenta) and 0069 (brown) plotted against photon energy.

Table 5.3 summarizes the counting efficiencies for the original models and all conducted registration simulations.

The lungs counting efficiency was calculated for energies in the region of interest (ROI) from $35keV$ to $80keV$. The counting efficiencies of each of the detectors is calculated, then the efficiencies of both detectors are summerized and referenced to that of the MEETMan for comparison.

The counting efficiency of the original ICRP adult male torso model is 51% higher than in the case of MEETMan. All 12 deformed models showed results closer to that of the MEETMan. The effect of volume control was minor, further testing must be conducted before clear conclusions can be derived.

Including the lung surface as additional information for the registration, leads to better results, since they result in a better registration and thus to a CWT similar to that of the MEETMan model. Less than 4% deviations of the MEEMMan results were observed.

Using the 12 circular areas as landmarks in the third set of registration simulations lead to results similar to those when including the lungs surfaces. Increasing the stiffness of the lungs tissue improved the registration.

Table 5.3 The counting efficiency for the MEETMan torso model, the ICRP adult torso male model and all the registered models. The counting efficiencies for each of the detectors, their sum and the sum relative to that of the MEETMan are listed in the table.

Registration ID	Phoswich 1 (counts/photon)	Phoswich 2 (counts/photon)	Sum (counts/photon)	Sum relative to MEETMan
MEETMan	0.0156248	0.0155758	0.0312006	1.000
ICRP	0.0221249	0.0249486	0.0472735	1.509
B000	0.0169356	0.0192565	0.0361921	1.160
B001	0.0165879	0.0190415	0.0356294	1.142
B002	0.0167627	0.0190823	0.0358450	1.149
0054	0.0140412	0.0166958	0.030737	0.985
0055	0.0137575	0.0163063	0.0300638	0.964
0056	0.0140440	0.0466454	0.0306894	0.984
0057	0.0164183	0.0191963	0.0356146	1.141
0058	0.0164543	0.0191192	0.0355735	1.140
0059	0.0163096	0.0190683	0.0353779	1.137
0069	0.0150627	0.0156868	0.0307495	0.986
0070	0.0146197	0.0153394	0.0299591	0.960
0071	0.0143486	0.0151543	0.0295029	0.946

5.5 Discussion

In this chapter an elastic 3D image registration method, that uses the ICP algorithm and the modified mass-spring system Adamss was presented. The tool based on this method was called Morphi.

Three implementations of the method using Morphi were demonstrated namely the validation of heart modeling simulations (Section 5.4.1), tracking of ventricular deformation (Section 5.4.2) and the generation of personalized torso models for detector efficiency calibration in virtual radiation protection scenarios (Section 5.4.3).

In the presented method, the registration is performed iteratively where in each iteration selected points T_c from the template image T are associated with selected points R_c from the reference image R and the Euclidean distance d between the corresponding points is used to move the points of the template T into the direction of their corresponding points of the reference R. The modified mass-spring system Adamss, a physical model of elasticity is used to control the registration process by giving the template elastic properties and thus guiding or regularizing the registration.

Unlike other elastic registration methods, where models of isotropic linear elasticity are employed, the elasticity model used in this work (Adamss) is able to simulate inhomogeneous, non-linear, anisotropic behavior. Furthermore, it allows inhomogeneous volume preservation and volume control during the image registration process.

In this context, the elastic properties assigned to the materials in the deformable template image should not be understood as physical material properties, but rather as relative geometrical relations. For example, if a material of the template image had anisotropic mechanical properties, where the material is two times stiffer in one axis as compared to another, it means the body is allowed to grow two times more into the direction of the first axis in comparison to the second axis during the elastic registration process, regardless of the physical material properties.

By remembering that the model of elasticity serves as a regularizer in the registration process, one can see that these features increase the control over the registration process and thus improve its quality. However, using a model of elasticity is not suitable for the registration of images undergoing large deformations. In those cases other regularizers could be used e.g. models for fluid dynamics [12].

In the development of this method a variant of the ICP algorithm was created. This variant is very similar to the original ICP and differs from it in the grouping points from T_c and R_c in several subsets, where correspondences can only be assigned between points from T_c and R_c belonging to the same subset. Each subset can have different weight when calculating the distance measure.

Many improvement could be made to refine the used ICP variant and get better registration outcomes.

For instance, the normal-space sampling strategy that selects a large number of template and reference point in areas of features [308].

Linear search method was used to perform the correspondences association task, making the ICP algorithm used of $O(k \cdot m \cdot n)$ time computations complexity in the worst case where m is the number points in T_c and n the number of points in R_c. k-d tree could be used to speedup the variant back to the complexity of the original ICP which is $O(k \cdot m \cdot \log(n))$ as mentioned in Section 5.2.3.3. Other nearest neighbor problem general solutions like "Elias" [325] could be used.

Greenspan *et al.* [329] proposed a novel nearest neighbor algorithm called (STCNN). Unlike the general solutions, this algorithm was developed specially to be used in ICP implementations. STCNN stands for Spherical Triangle Constraint Nearest Neighbor, and, as the name implies, is based on the spherical and triangular constraints that reduce the number of possible correspondences for any given point of the source points, thus reducing the nearest neighbor search time. A comparison between the general solutions like k-d tree, "Elias" and the STCNN was conducted. The results shows that "Elias" performs much better than the k-d tree data structure method, and that STCNN performs even better.

D. Kim *et al.* [330] proposed a method that consists of two acceleration techniques, hierarchical model point selection (HMPS), and logarithmic data point search (LDPS). The hierarchial model point selection assumes that closest reference points corresponding to two spatially close template points are also close to each other. And therefore, once the template points and their corresponding closest reference points are known, the search through all the template points is not needed. In other words, HMPS reduces the size of the search list and thus speeds up the algorithm. Logarithmic data point search (LDPS) assumes that the data points are smoothly and continuously distributed, a 2D logarithmic search method that does not scan all data points to find a closest reference points corresponding to a template point is used resulting in a further reduction of search time. With this method 3.17 speed up factor was achieved in comparison with ICP that uses the k-d tree algorithm for computing the nearest neighbor.

These algorithms, specially developed to be used with ICP implementations, could be used to speed up finding the correspondences.

However, it is important to mention that in the case of Morphi, the time needed for the minimization of the cost function E in Eq. (5.10) is the bottleneck of the registration method.

A method using Morphi to validate heart modeling simulations was presented in Section 5.4.1 where the deformed model resulting from the simulation at systole, the template T, is registered onto a segmented image of the ventricles at systole which is in turn the reference R.

The method selects points from the surfaces of the ventricles in the template T and the reference R images to generate the point clouds T_c and R_c. Landmarks defined using tagging points extracted from a FLASH MRI sequence could be included into the registration process.

A large number of tagging landmarks that deliver detailed information on the movement of the ventricles should be available to ensure the quality of the validation. In case no tagging landmarks or few tagging landmarks are available, as in the case of the conducted simulations in Section 5.4.1, the results should be interpreted as a measure of similarity between the external shape of the template and that of the reference.

Beside the tagging landmarks, adding landmarks that define anatomical features on the surfaces of the ventricles, as the apex and the valves would improve the registration. Increasing the stiffness used for the myocardial tissue, will force points to follow the landmarks more closely improving the registration locally around the landmarks, and thus globally on the organ level.

Because, every heart beat is different, one must be careful when validating the modeling by comparing computer simulations to a real heart beat. To ensure the comparison contains statistical deviations, the registration must be repeated for several reference images, which involves registration of several images on the ventricles taken at different times.

In Section 5.4.2, a method for tracking ventricular deformation was presented where a segmented dataset of the heart based on 4D MRI dataset is used as the template image T, and several segmented datasets of the heart in contracted states are used as references $R_i(i = 1, \ldots, n)$.

Creating the template points cloud T_c and reference points clouds $R_{c,i}$, the surfaces of the ventricles are used, similar to the method developed for the validation of ventricular modeling. Here also landmarks selected from tagging points extracted from a FLASH MRI sequence could be included into the registration process.

Since this method requires the segmentation of datasets in different contraction states, and the extraction of tagging points, it is very labour intensive and time consuming. Additionally, the accuracy of the registration is limited by the fidelity of the manual segmentation. Here too, using a large number of landmarks will significantly improve the quality of the registration and thus the task of tracking the ventricular deformation.

Section 5.4.3 presents a method that uses Morphi to generate personalized torso models for photon detectors efficiency calibration.

The personalized torso is generated by registration of the surface of a segmented torso, the template T, onto the surface scan of the individual's torso, the reference R. By assigning different stiffness curves to the tissues making up the already segmented torso, the registration can be controlled.

By measuring the individual's lungs volume and liver size, the volume control feature can be used to add additional information to the registration process.

The chest wall thickness (CWT) is a crucial value for the efficiency calibration of detectors. It can be approximated in several locations using routine ultrasound measurement, which can be translated to landmarks in both the template and the reference models. Features on the torso can be also interpreted as landmarks that can be used to improve the generation of personalized torsos.

Using anisotropic properties for the organs in the already segmented model, can result in a better registration. For examples, ribs could be assigned a lower stiffness in the longitudinal axis than in the transverse plane. Once again, it is important to remember that the stiffnesses in the case of this application of Morphi represent relative geometrical relations and not physical material properties.

The conducted simulations show that the method is able to generate personalized torsos good enough for the efficiency calibration of detectors. However, further experiments should be conducted on different male and female models ranging in age and shape, to obtain the right set of parameters for each case.

Volume control proved to have a minor effect on the outcome of the efficiency calibration, but further testing must be conducted before clear conclusions can be derived.

Differences of the counting efficiencies of the ICRP adult male and its derived models were observed for Phoswich 1 and 2. This is a result of the asymmetry of the lungs. This was not observed for the MEETMan model. The registration technique is obviously not able to compensate the left-right asymmetry. Therefore, single lung lobes could be investigated and treated individually.

Chapter 6
Outlook

It is almost a deterministic reality that collective scientific research goes only forwards. To help in that direction, some recommendations and suggestions about further developments are presented in this chapter.

An interesting subject to start with, is bridging the gap between the method of the virtual hexahedron described in Section 2.4.2.2 and FEM. Already, the mathematics behind the method of the virtual hexahedron and the distribution of forces to the particles using the method of Bourguignon give the impression that a mathematical link between both methods might be only few mathematical steps aways. In that case, the method will be equivalent to FEM. If not, it will be possible to quantify the differences between both methods, thus quantify the approximation made by using Adamss using that method.

Adamss performs many vector operation task in each time step. The use of GPUs for the processing might decrease the calculation time significantly, paving the road to model bigger objects or objects with higher resolution with the same computation time.

Using Adamss for modeling the elastomechanical behavior of tissues such as the heart, the female breast, the liver or skeletal muscles is possible. However most of these tissues is in direct contact with other tissues where sliding and friction occurs. Implementing a collision detection and a collision reaction model is crucial for such applications. Such an implementation would open a hole new range of applications where this method could play an important role. However the implemented method must be cost effective, otherwise it won't be able to compete against other well-established methods.

Until now, computational modeling fails in reproducing the behavior of the heart with high accuracy. Therefore, the focus must be set on developing a mathematical and computational method able to reproduce the deformation of the cardiac tissue during a heart cycle with high fidelity. The model should also be able to simulate different loading states of the heart. Eventually, parameters to personalize that model must be defined and the model must reproduce pathological behavior accurately.

Once that computational method is obtained. It is only then possible to understand the effect of neglecting one aspect of the modeling or another. That will pave the way to the development of simplified modeling methods with known deviation from reality. And engineers could then take conscious decisions about the method they need to implement the system to ensure a specific accuracy. Alternatively, they will be able to accurately calculate the error they make when using methods limiting the computational costs.

Adamss might be helpful in its current form to investigate hypotheses such as the contraction of the heart in case the force development model output is doubled in amplitude. Or the relaxation behavior of the heart in case the tension development curves were narrower. But, using Adamss for very accurate modeling heart elastomechanics is not highly recommended. Especially that better modeling will involve implementing blood pressure, blood flow and pericardium models including the collision detection and collision reaction models. Well-established methods in this phase are expected to provide better insights toward better modeling.

On the contrary, because of its flexibility using the modified mass-spring system for 3D elastic image registration as described in Chapter 5 is recommended, because of the control it can offer to the registration process. Other implementations for the elastic registration can be thought of, such as automatic registration of the heart chamber, or the automatic full body segmentation of men, women and children belonging to different age groups.

*****

There are many possibilities of how further research will be carried on. However, one thing is sure "The flow is in the human nature... that can't stop" and so is scientific research. That might sounds utopian but "A map of the world that does not include Utopia is not worth even glancing at, for it leaves out the one country at which Humanity is always landing".

References

1. T. Belytschko, W. Kam Liu, and B. Moran, *Nonlinear Finite Elements for Continua and Structures*. Wiley, 2000.
2. F. B. Sachse, *Computational cardiology : modelling of anatomy, electrophysiology, and mechanics*, vol. 2966 of *Lecture Notes in Computer Science*. Heidelberg: Springer, 2004.
3. Y. C. Fung, *Foundations of Solid Mechanics*. Englewood Cliffs, New Jersey: Prentice Hall, 1965.
4. A. E. Green and W. Zerna, *Theoretical elasticity*. New York: Dover, 1992.
5. R. W. Ogden, *Non-linear elastic deformations*. Dover books on physics, Mineola, NY: Dover Publ., 1997.
6. C. Oomens, M. Brekelmans, and F. Baaijens, *Biomechanics: Concepts and Computation*. Cambridge Texts in Biomedical Engineering, Cambridge University Press, 2009.
7. M. Nash and University of Auckland, Dep. of Engineering Science, New Zealand, *Mechanics and Material Properties of the Heart using an Anatomically Accurate Mathematical Model*. PhD thesis, 1998.
8. S. F. F. Gibson and B. Mirtich, "A Survey of Deformable Modeling in Computer Graphics," *Merl*, 1997.
9. A. Nealen, M. Muller, R. Keiser, E. Boxerman, and M. Carlson, "Physically based deformable models in computer graphics," *Computer graphics forum*, vol. 25, pp. 809–836, 2006.
10. P. Suetens, *Fundamentals of Medical Imaging*. Cambridge University Press, 2002.
11. F. B. Sachse, C. Werner, M. Müller, and K. Meyer-Waarden, "Segmentation and tissue classification of the Visible Man dataset using the computertomographic scans and the thin-section photos," *Proc. First Users Conference of the National Library of Medicine*, pp. 125–126, 1996.
12. J. Modersitzki, *Numerical Methods for Image Registration*. New York: Oxford University Press, 2004.
13. J. Burney, "Echocardiogram at the 47th Combat Support Hospital, US Army, TIKRIT, Iraq," Sept. 2009.
14. K. Lenes, "Echocardiogram. Ultrasound picture of the heart in four chamber viev. Left side of the heart to the right, apex down," May 2005.
15. O. Dössel, *Bildgebende Verfahren in der Medizin*. Berlin, Heidelberg, New York: Springer, 2000.
16. P. Suetens, *Fundamentals of medical imaging*. Cambridge medicine, Cambridge: Cambridge Univ. Pr., 2009.
17. E. Keeve, S. Girod, and B. Girod, "Craniofacial surgery simulation," in *Proceedings of the 4th International Conference on Visualization in Biomedical Computing VBC*, 1998.
18. G. Debunne, M. Desbrun, A. H. Barr, and M. P. Cani, "Interactive multiresolution animation of deformable models," in *Eurographics Workshop on Computer Animation and Simulation*, 1999.
19. F. B. d. Casson and C. Laugier, "Modeling the dynamics of a human liver for a minimally invasive surgery simulator," in *Proceedings of the Second International Conference on Medical Image Computing and Computer-Assisted Intervention (MICCAI*, vol. Lecture Notes In Computer Science, Vol. 1678, pp. 1156–1165, 1999.

20. J. Mosegaard, P. Herborg, and T. S. Sörensen, "A GPU accelerated spring mass system for surgical simulation," *Studies in health technology and informatics*, vol. 111, pp. 342–348., 2005.

21. F. Dogan and Y. Atilgan, "Physically based Virtual Surgery Planning and Simulation Tools for Personal Health Care Systems," *1st International ICST Symposium on IT Revolutions*, 2008.

22. H. Delingette, "Towards realistic soft tissue modeling in medical simulation," tech. rep., 1998.

23. M. P. Nash and P. J. Hunter, "Computational mechanics of the heart. From tissue structure to ventricular function," *Journal of Elasticity*, vol. 61, pp. 113–141, 2000.

24. R. C. P. Kerckhoffs, M. L. Neal, Q. Gu, J. B. Bassingthwaighte, J. H. Omens, and A. D. McCulloch, "Coupling of a 3D finite element model of cardiac ventricular mechanics to lumped systems models of the systemic and pulmonic circulation," *Annals of Biomedical Engineering*, vol. 35, pp. 1–18, 2007.

25. V. Rajagopal, J. H. Chung, D. P. Bullivant, P. M. F. Nielsen, and M. P. Nash, "Determining the finite elasticity reference state from a loaded configuration," *International Journal for Numerical Methods in Engineering*, vol. 72, pp. 1434–1451, 2007.

26. M. Sermesant, F. Billet, R. Chabinjok, T. Mansi, P. Chinchapatnam, P. Moireau, J.-M. Peyrat, K. Rhode, M. Ginks, P. Lambiase, S. Arridge, H. Delingette, M. Sorine, C. A. Rinaldi, D. Chapelle, R. Razavi, and N. Ayache, "Personaises electromechanical model of the heart for the prediction of the acute effects of cardiac resynchronisation therapy," in *Functional Imaging and Modeling of the Heart 2009*, 2009.

27. P. Volino and N. M. Thalmann, "Developing simulation techniques for an interactive clothing system," in *Virtual Systems and MultiMedia, 1997. VSMM*, pp. 109–118, 1997.

28. M. Desbrun and M. P. Cani-Gascuel, "Active Implicit Surface for Animation," in *Graphics Interface*, pp. 143–150, 1998.

29. D. Baraff and A. Witkin, "Large steps in cloth simulation," in *SIGGRAPH '98: Proceedings of the 25th annual conference on Computer graphics and interactive techniques*, (New York, NY, USA), pp. 43–54, Acm, 1998.

30. M. Desbrun, P. Schröder, and A. Barr, "Interactive animation of structured deformable objects," in *Proceedings of the 1999 conference on Graphics interface '99*, (San Francisco, CA, USA), pp. 1–8, Morgan Kaufmann Publishers Inc., 1999.

31. Y. M. Kang, J. H. Choi, H. G. Cho, D. H. Lee, and C. J. Park, "Real-time animation technique for flexible and thin objects," in *Proc. of WSCG 2000, Pizen, Czech.*, pp. 322–329, 2000.

32. Y. M. Kang, J. H. Choi, H. G. Cho, and C. J. Park, "Fast and stable animation of cloth with an approximated implicit method," in *CGI*, p. 247, IEEE Computer Society, Washington, DC, USA, 2000.

33. L. Li and V. Volkov, "Cloth animation with adaptively refined meshes," in *Proceedings of the Twenty-eighth Australasian conference on Computer Science - Volume 38*, Acsc, (Darlinghurst, Australia, Australia), pp. 107–113, Australian Computer Society, Inc., 2005.

34. K. Waters and D. Terzopoulos, "A physical model of facial tissue and muscle articulation," in *Proceedings of the 1st Conference on Visualization in Biomedical Computing*, pp. 77–82, IEEE, 1990.

35. L. P. Nedel and D. Thalmann, "Real time muscle deformations using mass-spring systems," in *Proceedings of the Computer Graphics International 1998* (F.-E. Wolter, ed.), (Washington, DC, USA), pp. 156–165, IEEE Computer Society, 1998.

36. Y. Chen, Q. Zhu, and A. Kaufman, "Physically-based animation of volumetric objects," in *Proc. IEEE Computer Animation*, pp. 154–160, 1998.

37. Q.-h. Zhu, Y. Chen, and A. Kaufman, "Real-time biomechanically-based muscle volume deformation using FEM," in *Proceedings of Endographics*, vol. 17, pp. 1–11, 1998.

38. D. Bourguignon and M.-P. Cani, "Controlling anisotropy in mass-spring systems," in *Computer Animation and Simulation*, (Interlaken), pp. 113–123, 2000.

39. M. Mohr, *A hybrid deformation model of ventricular myocardium*. PhD thesis, Karlsruhe, 2006.

40. J. Gross, M. Hauth, B. Eberhard, O. Etzmuss, L. Schnieder, and G. Buess, "Physically based tissue models for surgery simulations," in *SMIT 12th International Conference*, p. 320, 2000.

41. J. Dornheim, L. Dornheim, B. Preim, I. Hertel, and G. Strauss, "Generation and Initialization of Stable 3D Mass-Spring Models for the Segmentation of the Thyroid Cartilage," *Pattern Recognition*,

vol. 4174, pp. 162–171, 2006.

42. T. Boehler and H. O. Peitgen, "Evaluation of image registration using a mass-spring model of the breast," in *IFMBE Proceedings*, vol. 25, pp. 2201–2204, Springer, 2009.

43. O. Deussen, L. Kobbelt, and P. Tucke, "Using simulated annealing to obtain good nodal approximations of deformable bodies," in *Proc. 6th Eurographics Workshop on Animation and Simulation*, pp. 30–43, Springer, 1995.

44. J. Louchet, X. Provot, and D. Crochemore, "Evolutionary identification of cloth animation models," in *Proc. 6th Eurographics Workshop on Animation and Simulation*, pp. 44–54, 1995.

45. G. S. P. Miller, "The motion dynamics of snakes and worms," in *International Conference on Computer Graphics and Interactive Techniques. Proceedings of the 15th annual conference on Computer graphics and interactive techniques*, pp. 169–173, Addison Wesley, 1988.

46. V. Ng-Thow-Hing and E. Fiume, "Interactive display and animation of b-spline solids as muscle shape primitives," in *Proceedings of the 8th Eurographics Workshop on Computer Animation and Simulation*, p. 81?97, Springer-Verlag, 1997.

47. Y. Lee, D. Terzopoulos, and K. Walters, "Realistic modeling for facial animation," in *SIGGRAPH*, pp. 55–62, ACM Press, New York, 1995.

48. D. Hutchinson, M. Preston, and T. Hewitt, "Adaptive Refinement for Mass-Spring Simulations," in *In 7th Eurographics Workshop on Animation and Simulation*, pp. 31–45, Springer-Verlag, 1996.

49. J. Bonet and R. D. Wood, *Nonlinear continuum mechanics for finite element analysis*. Cambridge University Press, 1997.

50. J. Grandy and Lawrence Livermore National Laboratory, "Efficient computation of volume of hexahedral cells," tech. rep., United States, 1997.

51. E. B. CGAL, *CGAL-3.2 User and Reference Manual*, 2006.

52. D. Bandyopadhyay and J. Snoeyink, "Almost-Delaunay simplices: Robust neighbor relations for imprecise 3D points using CGAL," *Computational Geometry*, vol. 38, pp. 4–15, 2007.

53. F. B. Usabiaga and D. Duque, "Applications of computational geometry to the molecular simulation of interfaces," *Phys. Rev. E*, vol. 79, p. 046709, 2009.

54. Z. Farbman, G. Hoffer, Y. Lipman, D. Cohen-Or, and D. Lischinski, "Coordinates for instant image cloning," *ACM Trans. Graph.*, vol. 28, pp. 67–1, 2009.

55. U. M. Ascher and L. R. Petzold, *Computer Methods for Ordinary Differential Equations and Differential-Algebraic Equations*. Philadelphia, PA, USA: Society for Industrial and Applied Mathematics, 1998.

56. A. Bultheel and R. Cools, *The birth of Numerical Analysis*. World Scientific Publishing Company, 2009.

57. A. Neumaier, *Introduction to Numerical Analysis*. Cambridge University Press, 2001.

58. J. R. Dormand, *Numerical Methods for Differential Equations A Computational Approach*. CRC-Press, 1996.

59. S. Balay, K. Buschelman, W. D. Gropp, D. Kaushik, M. G. Knepley, L. C. McInnes, B. F. Smith, and H. Zhang, "PETSc Web page," 2001.

60. W. Schroeder, *The Visualization Toolkit*. Kitware Inc., 2003.

61. P. Hunter, M. P. Nash, and G. P. Sands, *Computational Biology of the Heart*, ch. Computational Electromechanics of the Heart, pp. 345–408. Chichester: John Wiley & Sons, 1997.

62. J. O. Hallquist, "LS-DYNA theory manual," *Livermore software Technology corporation*, 2006.

63. J. Ferlay, H.-R. Shin, F. Bray, D. Forman, C. Mathers, and D. M. Parkin, "Estimates of worldwide burden of cancer in 2008: GLOBOCAN 2008," *International Journal of Cancer*, vol. 127, pp. 2893–2917, 2010.

64. C. Mathers, D. M. Fat, J. T. Boerma, and World Health Organization, "The global burden of disease : 2004 update," tech. rep., Geneva, Switzerland :, 2008.

65. B. O. Anderson and S. R. Distelhorst, "Guidelines for International Breast Health and Cancer Control–Implementation. Introduction," 2008.

66. K. K. Hunt, G. L. Robb, E. A. Strom, and N. T. Ueno, *Breast Cancer*. M. D. Anderson Cancer Care Series, New York: Springer, 2008.

67. P. T. Huynh, A. M. Jarolimek, and S. Daye, "The false-negative mammogram," *Radiographics : a Review Publication of the Radiological Society of North America, Inc*, vol. 18, pp. 1137–54; quiz 1243, 1998.

68. S. W. Fletcher and J. G. Elmore, "Mammographic Screening for Breast Cancer," *New England Journal of Medicine*, vol. 348, pp. Massachusetts Medical Society–1680, 2003.

69. V. P. Jackson, R. E. Hendrick, S. A. Feig, and D. B. Kopans, "Imaging of the radiographically dense breast," *Radiology*, vol. 188, pp. 297–301, 1993.

70. M. B. Laya, E. B. Larson, S. H. Taplin, and E. White, "Effect of estrogen replacement therapy on the specificity and sensitivity of screening mammography," *Journal of the National Cancer Institute*, vol. 88, pp. 643–649, 1996.

71. R. D. Rosenberg, W. C. Hunt, M. R. Williamson, F. D. Gilliland, P. W. Wiest, C. A. Kelsey, C. R. Key, and M. N. Linver, "Effects of age, breast density, ethnicity, and estrogen replacement therapy on screening mammographic sensitivity and cancer stage at diagnosis: review of 183,134 screening mammograms in Albuquerque, New Mexico," *Radiology*, vol. 209, pp. 511–518, 1998.

72. J. P. Kosters and P. C. Gotzsche, "Regular self-examination or clinical examination for early detection of breast cancer," *Cochrane Database of Systematic Reviews (Online)*, p. Cd003373, 2003.

73. J. H. Chung, "Modelling Mammographic Mechanics," 2008.

74. A. T. Stavros, *Breast Ultrasound*. Lippincott Williams & Wilkins, 2003.

75. E. C. Fear, "Microwave imaging of the breast," *Technology in Cancer Research & Treatment*, vol. 4, pp. 69–82, 2005.

76. P. M. Meaney, M. W. Fanning, D. Li, S. P. Poplack, and K. D. Paulsen, "A clinical prototype for active microwave imaging of the breast," *Microwave Theory and Techniques, IEEE Transactions on*, vol. 48, pp. 1841–1853, 2000.

77. M. Lazebnik, L. McCartney, D. Popovic, C. B. Watkins, M. J. Lindstrom, J. Harter, S. Sewall, A. Magliocco, J. H. Booske, M. Okoniewski, and S. C. Hagness, "A large-scale study of the ultrawideband microwave dielectric properties of normal breast tissue obtained from reduction surgeries," *Physics in Medicine and Biology*, vol. 52, pp. 2637–2656, 2007.

78. E. Zastrow, S. K. Davis, M. Lazebnik, F. Kelcz, B. D. Van Veen, and S. C. Hagness, "Development of anatomically realistic numerical breast phantoms with accurate dielectric properties for modeling microwave interactions with the human breast," *IEEE Transactions on bio-Medical Engineering*, vol. 55, pp. 2792–2800, 2008.

79. P. M. Meaney, S. A. Pendergrass, M. W. Fanning, and K. D. Paulsen, "Importance of Using a Reduced Contrast Coupling Medium in 2D Microwave Breast ImaginG," *Journal of Electromagnetic Waves and Applications*, vol. 17, pp. 333–355, 2003.

80. A. Lazaro, D. Girbau, and R. Villarino, "Simulated and experimental investigation of microwave imaging using UWB," *Progress In Electromagnetics Research*, vol. 94, pp. 263–280, 2009.

81. S. C. Hagness, A. Taflove, and J. E. Bridges, "Three-dimensional FDTD analysis of a pulsed microwave confocal system for breast cancer detection: design of an antenna-array element," *Antennas and Propagation, IEEE Transactions on*, vol. 47, pp. 783–791, 1999.

82. X. Li, S. K. Davis, S. C. Hagness, D. W. van der Weide, and B. D. Van Veen, "Microwave imaging via space-time beamforming: experimental investigation of tumor detection in multilayer breast phantoms," *Microwave Theory and Techniques, IEEE Transactions on*, vol. 52, pp. 1856–1865, 2004.

83. H. Madjar and E. Mendelson, *The Practice of Breast Ultrasound: Techniques, Findings, Differential Diagnosis*. Thieme, 2009.

84. A. Cooper, *On the anatomy of the breast*. London: Longman, 1840.

85. D. T. Ramsay, J. C. Kent, R. A. Hartmann, and P. E. Hartmann, "Anatomy of the lactating human breast redefined with ultrasound imaging," *Journal of Anatomy*, vol. 206, pp. 525–534, 2005.

86. F. H. Netter, *Atlas der Anatomie des Menschen*. Stuttgart; New York: Thieme, 1997.

87. P. L. Williams, R. Warwick, M. Dyson, and L. H. Bannister, *Gray's anatomy of the human body*. 1995.

88. D. F. Klein, A. M. Skrobala, and R. S. Garfinkel, "Preliminary look at the effects of pregnancy on the course of panic disorder," *Anxiety*, vol. 1, pp. 227–232, 1994.

89. J. C. Heggie, "Survey of doses in screening mammography," *Australasian Physical & Engineering Sciences in Medicine / Supported by the Australasian College of Physical Scientists in Medicine and the Australasian Association of Physical Sciences in Medicine*, vol. 19, pp. 207–216, 1996.

90. N. Jamal, K.-H. Ng, D. McLean, L.-M. Looi, and F. Moosa, "Mammographic breast glandularity in Malaysian women: data derived from radiography," *AJR. American Journal of Roentgenology*, vol. 182, pp. 713–717, 2004.

91. D. B. Kopans, *Breast Imaging*. Lippincott Williams & Wilkins, 2006.

92. Y. C. Fung, *Biomechanics: Mechanical Properties of Living Tissues*. New York, Berlin, Heidelberg: Springer, 1993.

93. T. A. Krouskop, T. M. Wheeler, F. Kallel, B. S. Garra, and T. Hall, "Elastic moduli of breast and prostate tissues under compression," *Ultrasonic Imaging*, vol. 20, pp. 260–274, 1998.

94. P. Wellman, R. Howe, E. Dalton, K. Kern, and Harvard Biorobotics Laboratory, "Breast tissue stiffness in compression is correlated to histological diagnosis," tech. rep., 1999.

95. E. E. W. Van Houten, M. M. Doyley, F. E. Kennedy, J. B. Weaver, and K. D. Paulsen, "Initial in vivo experience with steady-state subzone-based MR elastography of the human breast," *Journal of Magnetic Resonance Imaging : JMRI*, vol. 17, pp. 72–85, 2003.

96. A. Samani, J. Zubovits, and D. Plewes, "Elastic moduli of normal and pathological human breast tissues: an inversion-technique-based investigation of 169 samples," *Physics in Medicine and Biology*, vol. 52, p. 1565, 2007.

97. A. Sarvazyan, D. Goukassian, E. Maevsky, and G. Oranskaja, "Elasticity imaging as a new modality of medical imaging for cancer detection," *Proc. Int. Workshop on Interaction of Ultrasound with Biological Media*, pp. 69–81, 1994.

98. S. A. Kruse, J. A. Smith, A. J. Lawrence, M. A. Dresner, A. Manduca, J. F. Greenleaf, and R. L. Ehman, "Tissue characterization using magnetic resonance elastography: preliminary results," *Physics in Medicine and Biology*, vol. 45, p. 1579, 2000.

99. R. Sinkus, M. Tanter, T. Xydeas, S. Catheline, J. Bercoff, and M. Fink, "Viscoelastic shear properties of in vivo breast lesions measured by MR elastography," *Magnetic Resonance Imaging*, vol. 23, pp. 159–165, 2005.

100. L. Han, J. A. Noble, and M. Burcher, "A novel ultrasound indentation system for measuring biomechanical properties of in vivo soft tissue," *Ultrasound in Medicine & Biology*, vol. 29, pp. 813–823, 2003.

101. V. Rajagopal, *Modelling breast tissue mechanics under gravity loading*. PhD thesis, 2007.

102. F. S. Azar, D. Metaxas, and M. Schnall, "A finite element model of the breast for predicting mechanical deformations during biopsy procedures," *Proceedings of the IEEE Workshop on Mathematical Methods in Biomedical Image Analysis*, pp. 38–45, 2000.

103. F. Azar, D. Metaxas, and M. Schnall, *Medical Image Computing and Computer-Assisted Intervention ? MICCAI 2001*, vol. 2208 of *Lecture Notes in Computer Science*, ch. Methods for Modeling and Predicting Mechanical Deformations of the Breast Under External Perturbations, pp. 1267–1270. Springer Berlin / Heidelberg, 2001.

104. A. Samani, J. Bishop, M. J. Yaffe, and D. B. Plewes, "Biomechanical 3-D finite element modeling of the human breast using MRI data," *IEEE Transactions on Medical Imaging*, vol. 20, pp. 271–279, 2001.

105. C. Zyganitidis, K. Bliznakova, and N. Pallikarakis, "A novel simulation algorithm for soft tissue compression," *Medical & Biological Engineering & Computing*, vol. 45, pp. 661–669, 2007.

106. J.-H. Chung, V. Rajagopal, P. M. F. Nielsen, and M. P. Nash, "Modelling mammographic compression of the breast," *Medical Image Computing and Computer-Assisted Intervention : MICCAI ... International Conference on Medical Image Computing and Computer-Assisted Intervention*, vol. 11, pp. 758–765, 2008.

107. P. Pathmanathan, D. J. Gavaghan, J. P. Whiteley, S. J. Chapman, and J. M. Brady, "Predicting tumor location by modeling the deformation of the breast," *IEEE Transactions on bio-Medical Engineering*, vol. 55, pp. 2471–2480, 2008.

108. N. V. Ruiter, T. O. Müller, R. Stotzka, H. Gemmeke, J. R. Reichenbach, and W. A. Kaiser, "Finite element simulation of the breast's deformation during mammography to generate a deformation model

for registration," in *Bildverarbeitung für die Medizin*, pp. 86–90, 2003.

109. A. P. del Palomar, B. Calvo, J. Herrero, J. Lopez, and M. Doblare, "A finite element model to accurately predict real deformations of the breast," *Medical Engineering & Physics*, vol. 30, pp. 1089–1097, 2008.

110. K. Bliznakova, Z. Bliznakov, V. Bravou, Z. Kolitsi, and N. Pallikarakis, "A three-dimensional breast software phantom for mammography simulation," *Physics in Medicine and Biology*, vol. 48, pp. 3699–3719, 2003.

111. L. Roose, W. Mollemans, D. Loeckx, F. Maes, and P. Suetens, "Biomechanically based elastic breast registration using mass tensor simulation," *Medical Image Computing and Computer-Assisted Intervention : MICCAI ... International Conference on Medical Image Computing and Computer-Assisted Intervention*, vol. 9, pp. 718–725, 2006.

112. N. V. Ruiter, R. Stotzka, T. O. Muller, H. Gemmeke, J. R. Reichenbach, and W. A. Kaiser, "Model-based registration of X-ray mammograms and MR images of the female breast," *Nuclear Science, IEEE Transactions on*, vol. 53, pp. 204–211, 2006.

113. S. Sarkar, Y. Zhang, Y. Qiu, D. B. Goldgof, and L. Li, "3D Finite Element Modeling of Nonrigid Breast Deformation for Feature Registration in X-ray and MR Images," in *Applications of Computer Vision, 2007. WACV*, p. 38, 2007.

114. C. Tanner, J. Hipwell, and D. Hawkes, *Digital Mammography*, vol. 5116 of *Lecture Notes in Computer Science*, ch. Statistical Deformation Models of Breast Compressions from Biomechanical Simulations, pp. 426–432. Springer Berlin / Heidelberg, 2008.

115. A. Samani and D. Plewes, "A method to measure the hyperelastic parameters of ex vivo breast tissue samples," *Physics in Medicine and Biology*, vol. 49, p. 4395, 2004.

116. C. Tanner, J. A. Schnabel, D. L. G. Hill, D. J. Hawkes, M. O. Leach, and D. R. Hose, "Factors influencing the accuracy of biomechanical breast models," *Medical Physics*, vol. 33, pp. 1758–1769, 2006.

117. V. Rajagopal, A. Lee, J.-H. Chung, R. Warren, R. P. Highnam, M. P. Nash, and P. M. F. Nielsen, "Creating individual-specific biomechanical models of the breast for medical image analysis," *Academic Radiology*, vol. 15, pp. 1425–1436, 2008.

118. V. Rajagopal, Y. Kvistedal, J. H. Chung, M. P. Nash, and P. M. F. Nielsen, "Modelling the skin-breast tissue interface," *Journal of Biomechanics*, vol. 39, p. 638, 2006.

119. V. Rajagopal, P. M. F. Nielsen, and M. P. Nash, "Modeling breast biomechanics for multi-modal image analysis - successes and challenges," *Wiley Interdisciplinary Reviews in Systems Biology*, 2009.

120. V. Rajagopal, M. Nash, R. Highnam, and P. Nielsen, *Digital Mammography*, vol. 5116 of *Lecture Notes in Computer Science*, ch. The Breast Biomechanics Reference State for Multi-modal Image Analysis, pp. 385–392. Springer Berlin / Heidelberg, 2008.

121. M. Lazebnik, D. Popovic, L. McCartney, C. B. Watkins, M. J. Lindstrom, J. Harter, S. Sewall, T. Ogilvie, A. Magliocco, T. M. Breslin, W. Temple, D. Mew, J. H. Booske, M. Okoniewski, and S. C. Hagness, "The UWCEM Numerical Breast Phantoms Repository," 2007.

122. M. Lazebnik, D. Popovic, L. McCartney, C. B. Watkins, M. J. Lindstrom, J. Harter, S. Sewall, T. Ogilvie, A. Magliocco, T. M. Breslin, W. Temple, D. Mew, J. H. Booske, M. Okoniewski, and S. C. Hagness, "A large-scale study of the ultrawideband microwave dielectric properties of normal, benign and malignant breast tissues obtained from cancer surgeries," *Physics in Medicine and Biology*, vol. 52, p. 6093, 2007.

123. A. C. o. Radiology, "Breast Imaging Reporting and Data System (BI-RADS) Atlas," tech. rep., Reston, 2003.

124. J. M. Guinebretiere, E. Menet, A. Tardivon, P. Cherel, and D. Vanel, "Normal and pathological breast, the histological basis," *European Journal of Radiology*, vol. 54, pp. 6–14, 2005.

125. J. Lorenzen, R. Sinkus, M. Biesterfeldt, and G. Adam, "Menstrual-cycle dependence of breast parenchyma elasticity: estimation with magnetic resonance elastography of breast tissue during the menstrual cycle," *Investigative Radiology*, vol. 38, pp. 236–240, 2003.

126. C. Williams and Ohio State University, *The biomechanics of normal breast tissue*. PhD thesis, 2000.

127. E. European Commission, "Cardiovascular Diseases - Medical Research - Health - Research - European Commission," Apr. 2011.

128. A. McCulloch, "Functionally and structurally integrated computational modeling of ventricular physiology," vol. 54, pp. 531–539, 2004.

129. F. W. Aelen, T. Arts, D. G. Sanders, G. R. Thelissen, A. M. Muijtjens, F. W. Prinzen, and R. S. Reneman, "Relation between torsion and cross-sectional area change in the human left ventricle," *Journal of Biomechanics*, vol. 30, pp. 207–212, 1997.

130. A. Van Der Toorn, P. Barenbrug, G. Snoep, F. H. Van Der Veen, T. Delhaas, F. W. Prinzen, J. Maessen, and T. Arts, "Transmural gradients of cardiac myofiber shortening in aortic valve stenosis patients using MRI tagging," *American Journal of Physiology. Heart and Circulatory Physiology*, vol. 283, pp. H1609–15, 2002.

131. R. Friedl, M. B. Preisack, W. Klas, T. Rose, S. Stracke, K. J. Quast, A. Hannekum, and O. Godje, "Virtual reality and 3D visualizations in heart surgery education," *The Heart Surgery Forum*, vol. 5, pp. E17–21, 2002.

132. M. Wierzbicki, M. Drangova, G. Guiraudon, and T. Peters, "Validation of dynamic heart models obtained using non-linear registration for virtual reality training, planning, and guidance of minimally invasive cardiac surgeries," *Medical Image Analysis*, vol. 8, pp. 387–401, 2004.

133. E. Coste-Maniere, L. Adhami, F. Mourgues, and A. Carpentier, "Planning, simulation, and augmented reality for robotic cardiac procedures: The STARS system of the ChIR team," *Seminars in Thoracic and Cardiovascular Surgery*, vol. 15, pp. 141–156, 2003.

134. J. M. Guccione, S. M. Moonly, A. W. Wallace, and M. B. Ratcliffe, "Residual stress produced by ventricular volume reduction surgery has little effect on ventricular function and mechanics: a finite element model study," *The Journal of Thoracic and Cardiovascular Surgery*, vol. 122, pp. 592–599, 2001.

135. J. M. Guccione, A. Salahieh, S. M. Moonly, J. Kortsmit, A. W. Wallace, and M. B. Ratcliffe, "Myosplint decreases wall stress without depressing function in the failing heart: a finite element model study," *The Annals of Thoracic Surgery*, vol. 76, pp. 1171–80; discussion 1180, 2003.

136. A. B. C. Dang, J. M. Guccione, P. Zhang, A. W. Wallace, R. C. Gorman, J. H. r. Gorman, and M. B. Ratcliffe, "Effect of ventricular size and patch stiffness in surgical anterior ventricular restoration: a finite element model study," *The Annals of Thoracic Surgery*, vol. 79, pp. 185–193, 2005.

137. A. Sambelashvili and I. R. Efimov, "Dynamics of virtual electrode-induced scroll-wave reentry in a 3D bidomain model," *American Journal of Physiology. Heart and Circulatory Physiology*, vol. 287, pp. H1570–81, 2004.

138. B. Rodriguez, B. M. Tice, J. C. Eason, F. Aguel, and N. Trayanova, "Cardiac vulnerability to electric shocks during phase 1A of acute global ischemia," *Heart Rhythm : the Official Journal of the Heart Rhythm Society*, vol. 1, pp. 695–703, 2004.

139. R. C. P. Kerckhoffs, P. H. M. Bovendeerd, F. W. Prinzen, K. Smits, and T. Arts, "Intra- and interventricular asynchrony of electromechanics in the ventricularly paced heart," *Journal of Engineering Mathematics*, vol. Volume 47, pp. 201–216, 2003.

140. R. C. P. Kerckhoffs, O. P. Faris, P. H. M. Bovendeerd, F. W. Prinzen, K. Smits, E. R. McVeigh, and T. Arts, "Electromechanics of paced left ventricle simulated by straightforward mathematical model: comparison with experiments," *American Journal of Physiology. Heart and Circulatory Physiology*, vol. 289, pp. H1889–97, 2005.

141. R. C. P. Kerckhoffs, P. H. M. Bovendeerd, J. C. S. Kotte, F. W. Prinzen, K. Smits, and T. Arts, "Homogeneity of cardiac contraction despite physiological asynchrony of depolarization: a model study," *Annals of Biomedical Engineering*, vol. 31, pp. 536–547, 2003.

142. R. C. P. Kerckhoffs, S. N. Healy, T. P. Usyk, and A. D. McCulloch, "Computational methods for cardiac electromechanics," *Proceedings of the IEEE*, vol. 94, pp. 769–783, 2006.

143. G. Pocock and C. D. Richards, *The Human Body: An introduction for the biomedical and health sciences*. Oxford: OUP, 2009.

144. L. Sherwood, *Human physiology : from cells to systems*. Australia ; United Kingdom: Thomson Brooks/Cole, 2004.

145. R. F. Schmidt and F. Lang, *Physiologie des Menschen*. Heidelberg: Springer Medizin Verlag, 2007.

146. R. M. Berne and M. N. Levy, *Cardiovascular physiology*. 1967.

147. G. J. Tortora and M. T. Nielsen, *Principles of Human Anatomy*. Wiley, 2009.

148. A. M. Katz, *Physiology of the Heart*, ch. 1. Structure of the heart and cardiac muscle, pp. 1–36. Raven Press, New York, 1992.

149. A. Vander, J. Sherman, and D. Luciano, *Human Physiology: The Mechanisms of Body Function*. W C B/McGraw-Hill, 1998.

150. K. T. Weber, Y. Sun, S. C. Tayagi, and P. M. Cleutjens, "Collagen network of the myocardium: function, structural remodeling and regulatory mechanisms," *Journal of Molecular and Cellular Cardiology*, vol. 26, pp. 279–292, 1994.

151. P. Hunter, B. H. Smaill, P. M. F. Nielsen, and I. J. L. Grice, *Computational Biology of the Heart*, ch. A Mathematical Model of Cardiac Anatomy, pp. 171–216. Chichester: John Wiley & Sons, 1997.

152. I. J. LeGrice, B. H. Smaill, L. Z. Chai, S. G. Edgar, J. B. Gavin, and P. J. Hunter, "Laminar structure of the heart: ventricle myocyte arrangement and connective tissue architecture in the dog," *American Journal of Physiology*, vol. 269, pp. H571–82, 1995.

153. M. Nash, *Mechanics and Material Properties of the Heart using an Anatomically Accurate Mathematical Model*. PhD thesis, 1998.

154. J. B. Caulfield and T. K. Borg, "The Collagen Network of the Heart," *Laboratory Investigation*, vol. 40, pp. 364–372, 1979.

155. D. D. Streeter, *Handbook of Physiology. Section 2: The Cardiovascular System*, vol. Vol. I. The Heart, ch. Gross morphology and fiber geometry of the heart, pp. 61–112. Baltimore: Williams and Wilkins: American Physiology Society, 1979.

156. A. A. Young, I. J. Legrice, M. A. Young, and B. H. Smaill, "Extended confocal microscopy of myocardial laminae and collagen network," *Journal of Microscopy*, vol. 192, pp. 139–150, 1998.

157. L. Zhukov and A. H. Barr, "Heart-muscle fiber reconstruction from diffusion tensor MRI," in *Proceedings of IEEE Visualization, 2003*, pp. 597–602, 2003.

158. S. Köhler, K. Hiller, C. Waller, P. M. Jakob, W. R. Bauer, and A. Haase, "Investigation of the Microstructure of the Isolated Rat Heart: A Comparison Between T*2- and Diffusion Weighted MRI," *Magnetic Resonance in Medicine*, vol. 50, pp. 1144–1150, 2003.

159. S. Y. Ho and D. Sanchez-Quintana, "The importance of atrial structure and fibers," *Clinical Anatomy (New York, N.Y.)*, vol. 22, pp. 52–63, 2008.

160. A. C. Guyton and J. E. Hall, *Textbook of Medical Physiology*. Saunders, 2000.

161. L. Sherwood, *Human physiology*. Thomson/Brooks/Cole, 2007.

162. D. W. Fawcett and N. S. McNutt, "The ultrastructure of cat myocardium. I. Ventricular papillary muscle," *Cell Biol.*, vol. 42, pp. 1–45, 1969.

163. J. Dudel, *Physiologie des Menschen*, ch. Grundlagen der Zellphysiologie, pp. 3–19. Berlin, Heidelberg, New York: Springer, 2000.

164. H. Lodish, A. Berk, P. Matsudaira, C. A. Kaiser, M. Krieger, M. P. Scott, S. L. Zipursky, and J. Darnell, *Molecular Cell Biology*. New York: W. H. Freeman and Company, 2003.

165. H. Lodish, A. Berk, S. L. Zipursky, P. Matsudaira, D. Baltimore, and J. Darnell, *Molecular Cell Biology*. New York: W. H. Freeman and Company, 1999.

166. D. M. Bers, *Excitation-contraction coupling and cardiac contractile force*, vol. 122 of *Developments in cardiovascular medicine*. Dordrecht: Kluwer, 1991.

167. J. Zhai, A. G. Schmidt, B. D. Hoit, Y. Kimura, D. H. MacLennan, and E. G. Kranias, "Cardiac-specific Overexpression of a Superinhibitory Pentameric Phospholamban Mutant Enhances Inhibition of Cardiac Function in Vivo," *J. Biol. Chem.*, vol. 275, pp. 10538–10544, 2000.

168. K. Haghighi, A. G. Schmidt, B. D. Hoit, A. G. Brittsan, A. Yatani, J. W. Lester, J. Zhai, Y. Kimura, G. W. D. II, D. H. MacLennan, and E. G. Kranias, "Superinhibition of sarcoplasmic reticulum function by phospholamban induces cardiac contractile failure," *J. Biol. Chem.*, vol. 276, pp. 24145–24152, 2001.

169. J. Malmivuo and R. Plonsey, *New York Oxford: Oxford University Press*, ch. "Nerve and Muscle Cells" in Bioelectromagnetism, pp. 34–43. 1995.

170. P. B. Bennett and H. G. Shin, *Cardiac Electrophysiology. From Cell to Bedside*, ch. Biophysics of Cardiac Sodium Channels, pp. 67–78. Philadelphia: W. B. Saunders Company, 1999.

171. M. C. Sanguinetti and M. Tristani-Firouzi, *Cardiac Electrophysiology. From Cell to Bedside*, ch. Delayed and Inward Rectifier Potassium Channels, pp. 79–86. Philadelphia: W. B. Saunders Company,

1999.

172. K. Shivkumar and J. N. Weiss, *Cardiac Electrophysiology. From Cell to Bedside*, ch. Adenosine Triphospate-Sensitive Potassium Channels, pp. 86–93. Philadelphia: W. B. Saunders Company, 1999.

173. M. Vassalle, H. Yu, and I. S. Cohen, *Cardiac Electrophysiology. From Cell to Bedside*, ch. Pacemaker Channels and Cardiac Automaticity, pp. 94–103. Philadelphia: W. B. Saunders Company, 1999.

174. U. Ravens, "Mechano-electric feedback and arrhythmias," *Prog. Biophys. Mol. Biol.*, vol. 82, pp. 255–266, 2003.

175. R. Klinke, H.-C. Pape, and S. Silbernagl, *Physiologie*. Stuttgart: Georg Thieme Verlag Stuttgart, 2005.

176. E. Marbán and G. F. Tomaselli, *Cardiac Electrophysiology. From Cell to Bedside*, ch. Molecular Biology of Sodium Channels, pp. 1–8. Philadelphia: W. B. Saunders Company, 1999.

177. A. L. Hodgkin and A. F. Huxley, "A quantitative description of membrane current and its application to conduction and excitation in nerve," *Journal of Physiology*, vol. 117, pp. 500–544, 1952.

178. R. L. Rasmusson, *Cardiac Electrophysiology. From Cell to Bedside*, ch. Pharmacology of Potassium Channels, pp. 156–167. Philadelphia: W. B. Saunders Company, 1999.

179. G. Y. Oudit, R. J. Ramirez, and P. H. Backx, *Cardiac Electrophysiology. From Cell to Bedside*, ch. Voltage-regulated Potassium Channels, pp. 19–32. Philadelphia: W. B. Saunders Company, 2004.

180. M. C. Sanguinetti and M. Tristani-Firouzi, *Cardiac Electrophysiology. From Cell to Bedside*, ch. Gating of Cardiac Delayed Rectifier K+ Channels, pp. 88–95. Philadelphia: W. B. Saunders Company, 2004.

181. S. J. Carroll, J. Kurokawa, and R. S. Kass, *Cardiac Electrophysiology. From Cell to Bedside*, ch. KCNQ1/KCNE1 Macromolecular Signaling Complex: Channel Microdomains and Human Disease, pp. 143–150. Philadelphia: W. B. Saunders Company, 2004.

182. C. W. Balke, E. Marbán, and B. O'Rourke, *Cardiac Electrophysiology. From Cell to Bedside*, ch. Calcium Channels: Structure, Function, and Regulation, pp. 8–21. Philadelphia: W. B. Saunders Company, 1999.

183. T. J. Kamp, Z. Zhou, S. Zhang, J. C. Makielski, and C. T. January, *Cardiac Electrophysiology. From Cell to Bedside*, ch. Pharmacology of L- and T-Type Calcium Channels in the Heart, pp. 141–156. Philadelphia: W. B. Saunders Company, 1999.

184. G. Seemann, *Modeling of electrophysiology and tension development in the human heart*. PhD thesis, Universitätsverlag Karlsruhe, 2005.

185. M. G. Villa Petroff, J. Palomeque, and A. R. Mattiazzi, "Na(+)-Ca2+ exchange function underlying contraction frequency inotropy in the cat myocardium," *J. Physiol.*, vol. 550, pp. 801–817, 2003.

186. M. J. A. v. Kempen, C. Fromaget, A. F. M. Moorman, and W. H. Lamers, "Spatial distribution of connexin43, the major cardiac gap junction protein, in the developing and adult rat heart," *Circ. Res.*, vol. 68, pp. 1638–1651, 1991.

187. J. E. Saffitz, D. L. Lerner, and K. A. Yamada, *Cardiac Electrophysiology. From Cell to Bedside*, ch. Gap Junction Distribution and Regulation in the Heart, pp. 181–191. Philadelphia: W. B. Saunders Company, 2004.

188. B. C. Knollmann and D. M. Roden, "A genetic framework for improving arrhythmia therapy," *Nature*, vol. 451, pp. 929–936, 2008.

189. R. F. Schmidt, *Physiologie kompakt*, ch. Erregungsphysiologie des Herzens, pp. 197–204. Berlin, Heidelberg, New York: Springer, 1999.

190. M. S. Spach, J. F. Heidlage, and P. C. Dolber, *Cardiac Electrophysiology. From Cell to Bedside*, ch. The dual nature of anisotropic discontinous conduction in the heart, pp. 213–222. Philadelphia: W. B. Saunders Company, 1999.

191. A. G. Kléber, V. G. Fast, and S. Rohr, *Cardiac Electrophysiology. From Cell to Bedside*, ch. Continuous and Discontinuous Propagation, pp. 213–222. Philadelphia: W. B. Saunders Company, 1999.

192. W. Koch, "Weitere Mitteilungen über den Sinusknoten des Herzens," *Verhandlungsband der Deutschen Pathologischen Gesellschaft*, vol. 13, pp. 85–92, 1909.

193. R. H. Anderson and S. Y. Ho, "The architecture of the sinus node, the atrioventricular conduction axis, and the internodal atrial myocardium," *Journal of Cardiovascular Electrophysiology*, vol. 9, pp. 1033–1048, 1998.

194. B. Biological Sciences, "Electrical conduction system of the heart," Sept. 2009.

195. S. Silbernagl and A. Despopoulos, *Taschenatlas der Physiologie*, ch. Nerv und Muskel, Arbeit, pp. 42–76. Stuttgart: Georg Thieme Verlag, 2001.

196. E. Carafoli, "Calcium signaling: A tale for all seasons," *Proc. Natl. Acad. Sci. USA*, vol. 99, pp. 1115–1122, 2002.

197. S. Silbernagl and A. Despopoulos, *Taschenatlas der Physiologie*, ch. Herz und Kreislauf, pp. 186–221. Stuttgart: Georg Thieme Verlag, 2001.

198. A. M. Gordon, E. Homsher, and M. Regnier, "Regulation of contraction in striated muscle," *Physiol. Rev.*, vol. 80, pp. 853–924, 2000.

199. D. M. Bers, "Cardiac excitation-contraction coupling," *Nature*, vol. 415, pp. 198–205, 2002.

200. D. P. Paolini, "Muscle."

201. E. H. Sonnenblick, "Series elastic and contractile elements in heart muscle: changes in muscle length," *Am J Physiol*, vol. 207, pp. 1330–1338, 1964.

202. J. G. Pinto and Y. C. Fung, "Mechanical properties of the heart muscle in the passive state," *J. Biomechanics*, vol. 6, pp. 597–616, 1973.

203. R. F. Janz, B. R. Kubert, and T. F. Moriarty, "Deformation of the diastolic left ventricle?II. Nonlinear geometric effects," *J. Biomechanics*, vol. 7, pp. 509–516, 1974.

204. N. R. Alpert, B. B. Hamrell, and W. Halpern, "Mechanical and biochemical correlates of cardiac hypertrophy," *Circulation Research*, vol. 35, pp. 71–82, 1974.

205. R. L. Kane, T. A. McMahon, R. L. Wagner, and W. H. Abelmann, "Ventricular elastic modulus as a function of age in syrian golden hamster," *Circulation Research*, vol. 38, pp. 74–80, 1976.

206. J. S. Rankin, C. E. Arentzen, P. A. McHale, D. Ling, and R. W. Anderson, "Viscoelastic properties of the diastolic left ventricle in the conscious dog," *Circulation Research*, vol. 41, pp. 37–45, 1977.

207. P. J. Hunter and B. H. Smaill, "The analysis of cardiac function: A continuum approach," *Prog. Biophys. Mol. Biol.*, vol. 52, pp. 101–164, 1988.

208. J. M. Guccione, *Theory of Heart*, ch. Finite element modeling of ventricular mechanics, pp. 121–144. Berlin, Heidelberg, New York: Springer, 1991.

209. V. P. Novak, F. C. P. Yin, and J. D. Humphrey, "Regional mechanical properties of passive myocardium," *J. Biomechanics*, vol. 27, pp. 403–412, 1994.

210. P. J. Hunter, *Molecular and subcellular cardiology: effects of structure and function*, vol. 30, ch. Myocardial constitutive laws for continuum mechanics models of the heart, pp. 303–318. New York: Plenum Press, 1995.

211. M. J. Moulton, L. L. Creswell, R. L. Actis, K. W. Myers, M. W. Vannier, B. A. Szabó, and M. K. Pasque, "An inverse approach to determining myocardial material properties," *J. Biomechanics*, vol. 28, pp. 935–948, 1995.

212. C. E. Miller, M. A. Vanni, and B. B. Keller, "Characterization of passive embryonic myocardium by quasi-linear viscoelasticity theory," *J. Biomechanics*, vol. 30, pp. 985–988, 1997.

213. J. M. Omens, S. M. Vaplon, B. Fazeli, and A. D. McCulloch, "Left ventricular geometric remodeling and residual stress in the rat heart," *Journal of Biomechanical Engineering*, vol. 120, pp. 715–719, 1998.

214. M. R. Zile, M. K. Cowles, J. M. Buckley, K. Richardson, B. A. Cowles, C. F. Baicu, G. Cooper, and V. Gharpuray, "Gel stretch method: A new method to measure constitutive properties of cardiac muscle cells," *Am. J. Physiol.*, vol. 274, p. 2188, 1998.

215. S. Dokos, I. J. LeGrice, B. H. Smaill, J. Kar, and A. A. Young, "A Triaxial-Measurement Shear-Test Device for Soft Biological Tissues," *J. Biomedical Engineering*, vol. 122, pp. 471–478, 2000.

216. R. J. Okamoto, M. J. Moulton, S. J. Peterson, D. Li, M. K. Pasque, and J. M. Guccione, "Epicardial Suction: A New Approach to Mechanical Testing of the Passive Ventricular Wall," *J. Biomed. Engin.*, vol. 122, pp. 479–487, 2000.

217. S. Dokos, B. H. Smaill, A. A. Young, and I. J. LeGrice, "Shear properties of passive ventricular myocardium," *Am. J. Physiol.*, vol. 283, p. 2650, 2003.

218. T. F. Robinson, M. A. Geraci, E. H. Sonnenblick, and S. M. Factor, "Coiled perimysial fibers of papillary muscle in rat heart: morphology, distribution, and changes in configuration," *Circulation research*, vol. 63, pp. 577–592, 1988.

219. J. S. Rankin, C. E. Arentyen, P. A. McHale, D. Ling, and R. W. Anderson, "Viscoelastic Properties of the Diastolic Left Ventricle in the Conscious Dog," *Circ. Res.*, vol. 41, pp. 37–45, 1976.

220. J. H. Omens and Y. C. Fung, "Residual strain in rat left ventricle," *Circulation Research*, vol. 66, pp. 37–45, 1990.

221. Y. C. Fung, *Biomechanics: Motion, Flow, Stress and Growth*. New York, Berlin, Heidelberg: Springer, 1990.

222. E. K. Rodriguez, J. H. Omens, L. K. Waldman, and A. D. McCulloch, "Effect of residual stress on transmural sarcomere length distributions in rat left ventricle," *The American Journal of Physiology*, vol. 264, pp. H1048–H1056, 1993.

223. L. A. Taber, N. Hu, T. Pexieder, E. B. Clark, and B. B. Keller, "Residual strain in the ventricle of the stage 16-24 chick embryo," *Circulation Research*, vol. 72, pp. 455–462, 1993.

224. K. D. Costa, K. May-Newman, D. Farr, and J. Ohmens, "Three-dimensional residual strain in mi-danterior canine left ventricle," *American Journal of Physiology-Heart and Circulatory Physiology*, vol. 273, pp. H1968–1976, 1997.

225. T. S. Harris, C. F. Baicu, C. H. Conrad, M. Koide, J. M. Buckley, M. Barnes, G. Cooper, and M. R. Zile, "Constitutive properties of hypertrophied myocardium: cellular contribution to changes in my-ocardial stiffness," *American Journal of Physiology- Heart and Circulatory Physiology*, vol. 282, pp. 2173–2182, 2002.

226. S. Silbernagl and A. Despopoulos, *Taschenbuch der Physiologie*. Stuttgart; New York: Georg Thieme Verlag, 1991.

227. S. Lundback, "Cardiac pumping and function of the ventricular septum," *Acta Physiologica Scandinavica. Supplementum*, vol. 550, pp. 1–101, 1986.

228. M. Carlsson, M. Ugander, H. Mosen, T. Buhre, and H. Arheden, "Atrioventricular plane displacement is the major contributor to left ventricular pumping in healthy adults, athletes, and patients with dilated cardiomyopathy," *American Journal of Physiology. Heart and Circulatory Physiology*, vol. 292, pp. H1452–9, 2007.

229. A. D. McCulloch, *The Biomedical Engineering Handbook*, ch. Cardiac biomechanics, pp. 28–1. CRC Press, 2000.

230. S. Chang, M. D. Cham, S. Hu, and Y. Wang, "3-T navigator parallel-imaging coronary MR angiography: targeted-volume versus whole-heart acquisition," *American Journal of Roentgenology*, vol. 191, pp. 38–42, 2008.

231. G. Seemann, D. U. J. Keller, D. L. Weiss, and O. Dössel, "Modeling Human Ventricular Geometry and Fiber Orientation Based on Diffusion Tensor MRI," in *Proc. Computers in Cardiology*, vol. 33, pp. 801–804, 2006.

232. R. Kalayciyan, D. Keller, G. Seemann, and O. Dössel, "Creation of a realistic endocardial stimulation profile for the visible man dataset," in *IFMBE Proceedings World Congress on Medical Physics and Biomedical Engineering*, vol. 25/4, pp. 934–937, 2009.

233. M. Lorange and R. M. Gulrajani, "A computer heart model incorporating anisotropic propagation. I. Model construction and simulation of normal activation," *Journal of Electrocardiology*, vol. 26, pp. 245–261, 1993.

234. P. Siregar, J. P. Sinteff, N. Julen, and B. Le Beux, "An interactive 3D anisotropic cellular automata model of the heart," *Computers and Biomedical Research*, vol. 31, pp. 323–347, 1998.

235. C. D. Werner, F. B. Sachse, and O. Dössel, "Electrical Excitation Propagation in the Human Heart," *Int. J. Bioelectromagnetism*, vol. 2, 2000.

236. K. Simelius, J. Nenonen, and B. M. Horacek, "Modeling Cardiac Ventricular Activation," *Int. J. Bioelectromagnetism*, vol. 3, 2001.

237. K. H. W. J. T. Tusscher and A. V. Panfilov, "Modelling of the ventricular conduction system," *Progress in Biophysics and Molecular Biology*, vol. 96, pp. 152–170, 2008.

238. D. U. J. Keller, R. Kalayciyan, O. Dössel, and G. Seemann, "Fast creation of endocardial stimulation profiles for the realistic simulation of body surface ECGs," in *IFMBE Proceedings World Congress on Medical Physics and Biomedical Engineering*, vol. 25/4, pp. 145–148, 2009.

239. D. Durrer, T. v. Dam, G. E. Freud, M. J. Janse, F. L. Meijler, and R. C. Arzbaecher, "Total excitation of the isolated human heart," *Circulation*, vol. 41, pp. 899–912, 1970.

240. D. Noble, "A modification of the Hodgkin-Huxley equations applicable to Purkinje fibre action and pace-maker potentials," *Journal of Physiology*, vol. 160, pp. 317–352, 1962.

241. K. H. W. J. t. Tusscher, D. Noble, P. J. Noble, and A. V. Panfilov, "A Model for Human Ventricular Tissue," *Am. J. Physiol.*, vol. 286, pp. H1573–H1589, 2004.

242. K. H. W. J. ten Tusscher and A. V. Panfilov, "Alternans and spiral breakup in a human ventricular tissue model," *American Journal of Physiology. Heart and Circulatory Physiology*, vol. 291, pp. H1088–100, 2006.

243. P. Wach, R. Killmann, F. Dienstl, and C. Eichtinger, "A computer model of human ventricular myocardium for simulation of ECG, MCG, and activation sequence including reentry rhythms," *Basic Research in Cardiology*, vol. 84, pp. 404–413, 1989.

244. M. Delorme, *Cellular Automata*, ch. An Introduction to Cellular Automata, pp. 5–49. Dordrecht: Kluwer, 1999.

245. R. A. FitzHugh, "Impulses and physiological states in theoretical models of nerve membran," *Biophysical Journal*, vol. 1, pp. 445–466, 1961.

246. J. P. Keener and A. V. Panfilov, *Computational Biology of the Heart*, ch. The effects of geometry and fibre orientation on propagation and extracellular potentials in myocardium, pp. 235–258. Chichester: John Wiley & Sons, 1997.

247. A. V. Panfilov, *Cardiac Electrophysiology. From Cell to Bedside*, ch. Three-dimensional wave propagation in mathematical models of ventricular fibrillation, pp. 271–277. Philadelphia: W. B. Saunders Company, 1999.

248. J. M. Rogers, "Modeling the Cardiac Action Potential Using B-Spline Surfaces," *IEEE Transactions on Biomedical Engineering*, vol. 47, pp. 784–791, 2000.

249. A. V. Hill, "The heat of shortening and the dynamic constants of muscle," *Proc. R. Soc. Lond. B*, vol. 126, pp. 136–195, 1938.

250. H. Chen, D. L. Weiss, G. Seemann, and O. Dössel, "Simulated Electromechanical Heterogeneity in Human Left Ventricle," in *Biomedizinische Technik*, vol. 50, pp. 564–565, 2005.

251. A. F. Huxley, "Muscle structures and theories of contraction," *Prog. Biophys. and biophys. Chemistry*, vol. 7, pp. 255–318, 1957.

252. R. B. Panerai, "A model of cardiac muscle mechanics and energetics," *Journal of Biomechanics*, vol. 13, pp. 929–940, 1980.

253. J. N. Peterson, W. C. Hunter, and M. R. Berman, "Estimated time course of Ca2+ bound to troponin C during relaxation in isolated cardiac muscle," *American Journal of Physiology*, vol. 260, pp. H1013–H1024, 1991.

254. A. Landesberg and S. Sideman, "Coupling calcium binding to troponin C and cross-bridge cycling in skinned cardiac cells," *Am J Physiol (Heart Circ Physiol 35)*, vol. 266, pp. H1260–H1271, 1994.

255. A. Landesberg and S. Sideman, "Mechanical regulation of cardiac muscle by coupling calcium kinetics with cross-bridge cycling: a dynamic model," *American Journal of Physiology*, vol. 267, pp. H779–H795, 1994.

256. P. J. Hunter, A. D. McCulloch, and H. E. D. J. t. Keurs, "Modeling the mechanical properties of cardiac muscle," *Progress in Biophysics & Molocular Biology*, vol. 69, pp. 289–331, 1998.

257. J. M. Guccione, I. Motabarzadeh, and G. I. Zahalak, "A distribution-moment model of deactivation in cardiac muscle," *J Biomech*, vol. 31, pp. 1069–1073, 1998.

258. J. J. Rice, R. L. Winslow, and W. C. Hunter, "Comparison of putative cooperative mechanisms in cardiac muscle: Length dependence and dynamic responses," *Am. J. Physiol.*, vol. 276, p. 1734, 1999.

259. J. J. Rice, M. S. Jafri, and R. L. Winslow, "Modeling Short-Term Interval-Force Relations in Cardiac Muscle," *Am. J. Physiol.*, vol. 278, p. 913, 2000.

260. M. Mlcek, J. Neumann, O. Kittnar, and V. Novak, "Mathematical model of the electromechanical heart contractile system–regulatory subsystem physiological considerations," *Physiol Res*, vol. 50, pp. 425–432, 2001.

261. D. P. Nickerson, N. P. Smith, and J. Hunter, "A model of cadiac cellular electromechanics," *Philosophical Transactions of the Royal Society London*, vol. 359, pp. 1159–1172, 2001.

262. K. Glänzel, F. B. Sachse, G. Seemann, C. Riedel, and O. Dössel, "Modeling force development in the sarcomere in consideration of electromechanical coupling," in *Biomedizinische Technik*, vol. 47-1/2,

pp. 774–777, 2002.

263. F. B. Sachse, K. Glänzel, and G. Seemann, "Modeling of Protein Interactions Involved in Cardiac Tension Development," *Int. J. Bifurc. Chaos*, vol. 13, pp. 3561–3578, 2003.

264. M. R. Wolff, S. H. Buck, S. W. Stoker, M. L. Greaser, and R. M. Mentzer, "Myofibrillar calcium sensitivity of isometric tension is increased in human dilated cardiomyopathies. Role of altered ß-adrenergically mediated protein phosphorylation," *Journal of Clinical Investigation*, vol. 98, pp. 167–176, 1996.

265. J. M. Guccione, *Theory of heart*, ch. Finite element modeling of ventricular mechanics, pp. 121–127. Springer-Verlag, 1991.

266. A. McCulloch, L. Waldmann, J. Rogers, and J. Guccione, "Large-scale finite element analysis of the beating heart," *Critical Reviews in Biomedical Engineering*, vol. 20, pp. 427–449, 1992.

267. J. Guccione, K. D. Costa, and A. D. McCulloch, "Finite element stress analysis of left ventricular mechanics in the beating dog heart," *Journal of Biomechanics*, vol. 28, pp. 1167–1177, 1995.

268. K. D. Costa, P. J. Hunter, J. S. Wayne, L. K. Waldman, J. M. Guccione, and A. D. McCulloch, "A three-dimensional finite element method for large elastic deformations of ventricular myocardium: II-prolate spheroidal coordinates," *Journal of Biomechanical Engineering*, vol. 118, pp. 464–472, 1996.

269. N. Ayache, D. Chapelle, F. Clement, Y. Coudiere, H. Delingette, J. A. Desideri, M. Sermesant, M. Sorine, and J. M. Urquiza, "Towards model-based estimation of the cardiac electro-mechanical activity from ECG signals and ultrasound images," *Lecture Notes in Computer Science*, vol. 2230, pp. 120–127, 2001.

270. N. P. Smith, P. J. Mulquiney, M. P. Nash, C. P. Bradley, D. P. Nickerson, and P. J. Hunter, "Mathematical modelling of the heart: cell to organ," *Chaos, Solitons & Fractal*, vol. 13, pp. 1613–1621, 2002.

271. M. Sermesant, H. Delingette, and N. Ayache, "An electromechanical model of the heart for image analysis and simulation," *IEEE Transactions on Medical Imaging*, vol. 25, pp. 612–625, 2006.

272. M. B. Mohr, L. G. Blümcke, G. Seemann, F. B. Sachse, and O. Dössel, "Volume modeling of myocard deformation with a spring mass system," in *Lecture Notes in Computer Science* (N. Ayache and H. Delingette, eds.), vol. 2673, (Berlin, Heidelberg, New York), pp. 332–339, Springer, 2003.

273. M. B. Mohr, R. R. Schnell, G. Seemann, F. B. Sachse, and O. Dössel, "Incorporating blood pressure load into an elastomechanical ventricular model," in *Computers in Cardiology, Lyon, France*, 2005.

274. J. M. Guccione, A. D. McCulloch, and L. K. Waldman, "Passive material properties of intact ventricular myocardium determined from a cylindrical model [see comments]," *J Biomech Eng*, vol. 113, pp. 42–55, 1991.

275. S. Balay, W. D. Gropp, L. C. McInnes, and B. F. Smith, "Efficient Management of Parallelism in Object Oriented Numerical Software Libraries," in *Modern Software Tools in Scientific Computing* (E. Arge, A. Bruaset, and H. Langtangen, eds.), pp. 163–202, Birkhäuser Press, 1997.

276. B. Sievers, S. Kirchberg, A. Bakan, U. Franken, and H.-J. Trappe, "Impact of papillary muscles in ventricular volume and ejection fraction assessment by cardiovascular magnetic resonance," *Journal of Cardiovascular Magnetic Resonance : Official Journal of the Society for Cardiovascular Magnetic Resonance*, vol. 6, pp. 9–16, 2004.

277. H. Schmid and The University of Auckland, *Passive myocardial mechanics : constitutive laws and materials parameter estimation*. PhD thesis, 2006.

278. C. S. Peskin and D. M. McQueen, "Fluid dynamics of the heart and its valves," *Case Studies in Mathematical Modeling: Ecology, Physiology, and Cell Biology*, 1997.

279. D. M. McQueen and C. S. Peskin, *Mechanics for a New Mellennium*, ch. Heart Simulation by an Immersed Boundary Method with Formal Second-order Accuracy and Reduced Numerical Viscosity, pp. 429–444. Springer Netherlands, 2002.

280. S. D. Mey, J. Vierendeels, and P. Verdonck, "Modeling of LV filling: A combined hydraulic and numerical approach," *Cardiovasc. Engin.*, vol. 1, pp. 163–169, 2001.

281. S. B. S. Krittian, *Modellierung der kardialen Strömung-Struktur-Wechselwirkung: Implicit coupling for KaHMo FSI*. PhD thesis, 2009.

282. P. Verdonck, P. Segers, and D. D. Mey, *Cardiac mechanics.* von Karman Institute for Fluid Dynamics, Brussel: VKI Lecture Series, 2003.

283. J. L. Weiss, J. W. Frederiksen, and M. L. Weisfeldt, "Hemodynamic determinants of the time-course of fall in canine left ventricular pressure," *J. Clin. Invest.*, vol. 58, pp. 751–760, 1979.

284. M. Levoy, K. Pulli, B. Curless, S. Rusinkiewicz, D. Koller, L. Pereira, M. Ginzton, S. Anderson, J. Davis, J. Ginsberg, J. Shade, and D. Fulk, "The Digital Michelangelo Project: 3D Scanning of Large Statues," in *Proceedings of ACM SIGGRAPH 2000*, pp. 131–144, 2000.

285. K. Ikeuchi, A. Nakazawa, K. Hasegawa, and T. Ohishi, "The Great Buddha Project: Modeling Cultural Heritage for VR Systems through Observation," in *ISMAR*, (Washington, DC, USA), p. 7, IEEE Computer Society, 2003.

286. D. Rey, G. Subsol, H. Delingette, and N. Ayache, "Automatic detection and segmentation of evolving processes in 3D medical images: Application to multiple sclerosis," *Medical Image Analysis*, vol. 6, pp. 163–179, 2002.

287. M. Ferrant, A. Nabavi, B. Macq, F. A. Jolesz, R. Kikinis, and S. K. Warfield, "Registration of 3-D intraoperative MR images of the brain using a finite-element biomechanical model," *IEEE Transactions on Medical Imaging*, vol. 20, pp. 1384–1397, 2001.

288. J. A. Schnabel, C. Tanner, A. D. Castellano-Smith, A. Degenhard, M. O. Leach, D. R. Hose, D. L. G. Hill, and D. J. Hawkes, "Validation of nonrigid image registration using finite-element methods: application to breast MR images," *IEEE Transactions on Medical Imaging*, vol. 22, pp. 238–247, 2003.

289. O. Clatz, H. Delingette, I.-F. Talos, A. J. Golby, R. Kikinis, F. A. Jolesz, N. Ayache, and S. K. Warfield, "Robust nonrigid registration to capture brain shift from intraoperative MRI," *IEEE Transactions on Medical Imaging*, vol. 24, pp. 1417–1427, 2005.

290. J. N. Yu, F. H. Fahey, H. D. Gage, C. G. Eades, B. A. Harkness, C. A. Pelizzari, and J. W. J. Keyes, "Intermodality, retrospective image registration in the thorax," *Journal of Nuclear Medicine : Official Publication, Society of Nuclear Medicine*, vol. 36, pp. 2333–2338, 1995.

291. Y.-C. Tai, K. P. Lin, C. K. Hoh, S. C. H. Huang, and E. J. Hoffman, "Utilization of 3-D elastic transformation in the registration of chest X-ray CT and whole body PET," in *Nuclear Science Symposium, 1996. Conference Record., 1996 IEEE*, vol. 3, pp. 1903–1907, 1996.

292. J. B. Maintz and M. A. Viergever, "A survey of medical image registration," *Medical Image Analysis*, vol. 2, pp. 1–36, 1998.

293. J. Cai, J. C. Chu, D. Recine, M. Sharma, C. Nguyen, R. Rodebaugh, V. A. Saxena, and A. Ali, "CT and PET lung image registration and fusion in radiotherapy treatment planning using the chamfer-matching method," *International Journal of Radiation Oncology, Biology, Physics*, vol. 43, pp. 883–891, 1999.

294. T. Mäkelä, S. Member, P. Clarysse, O. Sipilä, N. Pauna, Q. C. Pham, T. Katila, I. E. Magnin, and L. L. Axis, "A Review of Cardiac Image Registration Methods," 2002.

295. P. Foroughi and P. Abolmaesumi, "Elastic registration of 3D ultrasound images," *Medical Image Computing and Computer-Assisted Intervention : MICCAI ... International Conference on Medical Image Computing and Computer-Assisted Intervention*, vol. 8, pp. 83–90, 2005.

296. B. Fischer and J. Modersitzki, "Ill-posed medicine?an introduction to image registration," *Inverse Problems*, vol. 24, p. 034008, 2008.

297. P. M. Thompson, C. Vidal, J. N. Giedd, P. Gochman, J. Blumenthal, R. Nicolson, A. W. Toga, and J. L. Rapoport, "Mapping adolescent brain change reveals dynamic wave of accelerated gray matter loss in very early-onset schizophrenia," *Proceedings of the National Academy of Sciences of the United States of America*, vol. 98, pp. 11650–11655, 2001.

298. V. Arsigny, P. Fillard, X. Pennec, and N. Ayache, "Log-Euclidean metrics for fast and simple calculus on diffusion tensors," *Magnetic Resonance in Medicine : Official Journal of the Society of Magnetic Resonance in Medicine / Society of Magnetic Resonance in Medicine*, vol. 56, pp. 411–421, 2006.

299. J. Dauguet, A. Dubois, A.-S. Herard, L. Besret, G. Bonvento, P. Hantraye, and T. Delzescaux, "Towards a routine analysis of anatomical and functional post mortem slices in 3 dimensions," *Journal of Cerebral Blood Flow & Metabolism*, vol. 25, pp. S559–S559, 2005.

300. S. Periaswamy and H. Farid, "Medical image registration with partial data," *Medical Image Analysis*, vol. 10, pp. 452–464, 2006.

301. J. Modersitzki, *Fair: Flexible Algorithms for Image Registration (Fundamentals of Algorithms)*. Society for Industrial and Applied Mathematics, 2009.

302. Y. Chen and G. Medioni, "Object modeling by registration of multiple range images," in *International Conference on Robotics and Automation*, vol. 3, pp. 2724–2729, IEEE, 1991.

303. P. J. Besl and H. D. Mckay, "A method for registration of 3-D shapes," *Pattern Analysis and Machine Intelligence, IEEE Transactions on*, vol. 14, pp. 239–256, 1992.

304. B. Zitova, "Image registration methods: a survey," *Image and Vision Computing*, vol. 21, pp. 977–1000, 2003.

305. B. K. P. Horn, "Closed-form solution of absolute orientation using unit quaternions," *Journal of the Optical Society of America A*, vol. 4, pp. 629–642, 1987.

306. B. K. P. Horn, H. M. Hilden, and S. Negahdaripour, "Closed-Form Solution of Absolute Orientation using Orthonormal Matrices," *Journal Of The Optical Society America*, vol. 5, pp. 1127–1135, 1988.

307. K. S. Arun, T. S. Huang, and S. d. Blostein, "Least-squares fitting of two 3-D point sets," *IEEE Transactions on Pattern Analysis and Machine Intelligence*, vol. 9, pp. 698–700, 1987.

308. S. Rusinkiewicz and M. Levoy, "Efficient Variants of the ICP Algorithm," in *Third International Conference on 3D Digital Imaging and Modeling (3DIM)*, 2001.

309. G. Turk and M. Levoy, "Zippered polygon meshes from range images," in *SIGGRAPH*, (New York, NY, USA), pp. 311–318, Acm, 1994.

310. T. Masuda, K. Sakaue, and N. Yokoya, "Registration and Integration of Multiple Range Images for 3-D Model Construction," *Pattern Recognition, International Conference on*, vol. 1, p. 879, 1996.

311. S. Weik, "Registration of 3-D Partial Surface Models Using Luminance and Depth Information," in *In Proc. Int. Conf. on Recent Advances in 3-D Digital Imaging and Modeling*, pp. 93–100, 1997.

312. D. Simon, *Fast and Accurate Shape-Based Registration*. PhD thesis, Pittsburgh, PA, 1996.

313. G. Blais and M. D. Levine, "Registering multiview range data to create 3D computer objects," *IEEE Transactions on Pattern Analysis and Machine Intelligence*, vol. 17, pp. 820–824, 1995.

314. P. J. Neugebauer, "Geometrical cloning of 3D objects via simultaneous registration of multiple range images," in *SMA*, (Washington, DC, USA), p. 130, IEEE Computer Society, 1997.

315. K. Pulli, *Surface Reconstruction and Display from Range and Color Data*. PhD thesis, 1997.

316. G. Godin, M. Rioux, and R. Baribeau, "Three-dimensional registration using range and intensity information," *Proceedings of SPIE. Videometrics III*, vol. 2350, pp. 279–290, 1994.

317. K. Pulli, "Multiview Registration for Large Data Sets," in *Proceedings of the Second International Conference on 3-D Digital Imaging and Modeling*, (Ottawa, ON, Canada), pp. 160–168, IEEE, 1999.

318. C. Dorai, G. Wang, A. K. Jain, and C. Mercer, "Registration and integration of multiple object views for 3D model construction," *IEEE Transactions on Pattern Analysis and Machine Intelligence*, vol. 20, pp. 83–89, 1998.

319. D. Chetverikov, D. Stepanov, and P. Krsek, "Robust Euclidean alignment of 3D point sets: the trimmed iterative closest point algorithm," *Image and Vision Computing*, vol. 23, pp. 299–309, 2005.

320. M. Walker, L. Shao, and R. Volz, "Estimating 3-D location parameters using dual number quaternions," *CVGIP: Image Understanding*, vol. 54, pp. 358–367, 1991.

321. A. W. Fitzgibbon, "Robust Registration of 2D and 3D Point Sets," in *In British Machine Vision Conference*, pp. 411–420, 2001.

322. P. J. Rousseeuw, "Least Median of Squares Regression," *Journal of the American Statistical Association*, vol. 79, pp. 871–880, 1984.

323. A. Almhdie, C. Léger, M. Deriche, and R. Lédée, "3D registration using a new implementation of the ICP algorithm based on a comprehensive lookup matrix: Application to medical imaging," *Pattern Recogn. Lett.*, vol. 28, pp. 1523–1533, 2007.

324. J. L. Bentley, "Multidimensional binary search trees used for associative searching," *Commun. ACM*, vol. 18, pp. 509–517, 1975.

325. R. Rivest, "the optimality of elias?s algorithm for performing best-match searches," 1974.

326. K. Fukunage and P. M. Narendra, "A branch and bound algorithm for computing k-nearest neighbors," *IEEE Transactions on Computers*, vol. C-24, pp. 750–753, 1975.

327. E. V. Ruiz, "An algorithm for finding nearest neighbours in (approximately) constant average time," *Pattern Recogn. Lett.*, vol. 4, pp. 145–157, 1986.

328. M. Greenspan, G. Godin, and J. Talbot, "Acceleration of binning nearest neighbor nethods," in *Proceedings of Vision Interface*, (Montréal, Québec, Canada), NRC Punblications, 2000.

329. M. Greenspan and G. Godin, "A nearest neighbor method for efficient ICP," *3D Digital Imaging and Modeling, International Conference on*, pp. 161–168, 2001.

330. D. Kim, "A fast ICP algorithm for 3-D human body motion tracking," *ignal Processing Letters, IEEE*, vol. 17, pp. 402–405, 2010.

331. P. Yan and K. W. Bowyer, "A fast algorithm for ICP-based 3D shape biometrics," *Comput. Vis. Image Underst.*, vol. 107, pp. 195–202, 2007.

332. J. Feldmar and N. Ayache, "Rigid, affine and locally affine registration of free-form surfaces," *International Journal of Computer Vision*, vol. 18, pp. 99–119, 1996.

333. J. Feldmar, J. Declerck, G. Malandain, N. Ayache, I. Sophia, P. Epidaure, and S. A. Cedex, "Extension of the ICP Algorithm to non rigid Intensity-based Registration of 3D Volumes," in *Computer Vision and Image Understanding*, pp. 84–93, 1996.

334. C. Tong, S.-i. Kamata, and A. Ahrary, "3D face recognition based on fast feature detection and non-rigid iterative Closest Point," vol. 4, pp. 509–512, 2009.

335. B. Amberg, S. Romdhani, and T. Vetter, "Optimal step nonrigid ICP algorithms for surface Registration," in *IEEE Conference on Computer Vision and Pattern Recognition*, pp. 1–8, 2007.

336. B. Allen, B. Curless, and Z. Popovic, "The Space of Human Body Shapes: reconstruction and parameterization from range scans," *ACM Trans. Graph*, vol. 22, pp. 587–594, 2003.

337. N. Hasler, C. Stoll, M. Sunkel, B. Rosenhahn, H.-p. Seidel, and M. Informatik, "A statistical model of human pose and body shape.," 2009.

338. M. Ferrant, S. K. Warfield, C. R. G. Guttmann, R. V. Mulkern, F. A. Jolesz, and R. Kikinis, "3D Image Matching Using a Finite Element Based Elastic Deformation Model," in *Medical Image Computing and Computer-Assisted Intervention ? MICCAI?99*, vol. 1679 of *Lecture Notes in Computer Science*, pp. 202–209, Springer Berlin / Heidelberg, 1999.

339. J. Crouch, S. Pizer, E. Chaney, Y.-C. Hu, G. Mageras, and M. Zaider, "Automated Finite-Element Analysis for Deformable Registration of Prostate Images," *Medical Imaging, IEEE Transactions on*, vol. 26, pp. 1379–1390, 2007.

340. Y. Xie, M. Xie, and L. Yang, "A New Non-rigid Image Registration Algorithm Using the Finite-Element Method," *Education Technology and Computer Science, International Workshop on*, vol. 1, pp. 206–210, 2010.

341. S. Balay, K. Buschelman, V. Eijkhout, W. D. Gropp, D. Kaushik, M. G. Knepley, L. C. McInnes, B. F. Smith, and H. Zhang, "PETSc Users Manual," tech. rep., 2004.

342. D. U. J. Keller, O. Jarrousse, T. Fritz, S. Ley, O. Dössel, and G. Seemann, "Impact of Physiological Ventricular Deformation on the Morphology of the T-Wave: A Hybrid, Static-Dynamic Approach," *IEEE Trans. Biomed. Eng.*, p. (accepted), 2011.

343. L. Xia, M. Huo, Q. Wei, F. Liu, and S. Crozier, "Analysis of cardiac ventricular wall motion based on a three-dimensional electromechanical biventricular model," *Physics in Medicine & Biology*, vol. 50, pp. 1901–1917, 2005.

344. L. Hegenbart, *Numerical Efficiency Calibration of In Vivo Measurement Systems*. PhD thesis, 2009.

345. P. ICRP, "Adult Reference Computational Phantoms. ICRP Publication 110," tech. rep., 2009.

346. L. Hegenbart and H. Gün, "Determination of chest wall thickness of anthropometric voxel models," in *Proceedings of the Third European IRPA Congress, Helsinki, Finland*, 2010.

Karlsruhe Transactions on Biomedical Engineering
(ISSN 1864-5933)

Karlsruhe Institute of Technology / Institute of Biomedical Engineering (Ed.)

Die Bände sind unter www.ksp.kit.edu als PDF frei verfügbar oder als Druckausgabe bestellbar.

Band 2 Matthias Reumann
Computer assisted optimisation on non-pharmacological treatment of congestive heart failure and supraventricular arrhythmia. 2007
ISBN 978-3-86644-122-4

Band 3 Antoun Khawaja
Automatic ECG analysis using principal component analysis and wavelet transformation. 2007
ISBN 978-3-86644-132-3

Band 4 Dmytro Farina
Forward and inverse problems of electrocardiography : clinical investigations. 2008
ISBN 978-3-86644-219-1

Band 5 Jörn Thiele
Optische und mechanische Messungen von elektrophysiologischen Vorgängen im Myokardgewebe. 2008
ISBN 978-3-86644-240-5

Band 6 Raz Miri
Computer assisted optimization of cardiac resynchronization therapy. 2009
ISBN 978-3-86644-360-0

Band 7 Frank Kreuder
2D-3D-Registrierung mit Parameterentkopplung für die Patientenlagerung in der Strahlentherapie. 2009
ISBN 978-3-86644-376-1

Band 8 Daniel Unholtz
Optische Oberflächensignalmessung mit Mikrolinsen-Detektoren für die Kleintierbildgebung. 2009
ISBN 978-3-86644-423-2

Band 9 Yuan Jiang
Solving the inverse problem of electrocardiography in a realistic environment. 2010
ISBN 978-3-86644-486-7

Karlsruhe Transactions on Biomedical Engineering
(ISSN 1864-5933)